7 Day Loan

Guide to the Wiring Regulations

Guide to the Wiring Regulations
17th Edition IEE Wiring Regulations (BS 7671: 2008)

Darrell Locke IEng MIEE ACIBSE

Electrical Contractors' Association

Representing the best in electrical
engineering and building services

in association with

Extracts from BS 7671: 2008 have been kindly provided by the Institution of Engineering
and Technology (IET) and extracts from other standards have been reproduced with
permission from British Standards Institution (BSI). Information and copies of standards
are available from BSI at http://www.bsonline.bsi-global.com

John Wiley & Sons, Ltd

Copyright © 2008 John Wiley & Sons Ltd, The Atrium, Southern Gate, Chichester,
West Sussex PO19 8SQ, England
Telephone (+44) 1243 779777

Email (for orders and customer service enquiries): cs-books@wiley.co.uk
Visit our Home Page on www.wiley.com
Reprinted January 2008, February 2008.

ECA is the trademark of the Electrical Contractors' Association.

The ECA is the UK's largest and leading trade association representing electrical, electronic, installation engineering and
building services companies.

Website www.eca.co.uk

Whilst every care has been taken to ensure the accuracy of the information in this book, neither the author or the ECA can
accept liability for any inaccuracies or omissions arising from the information provided.

SELECT are Scotland's trade association for the electrical, electronics and communications systems industry.

Website www.select.org.uk

Other Wiley Editorial Offices

John Wiley & Sons Inc., 111 River Street, Hoboken, NJ 07030, USA

Jossey-Bass, 989 Market Street, San Francisco, CA 94103-1741, USA

Wiley-VCH Verlag GmbH, Boschstr. 12, D-69469 Weinheim, Germany

John Wiley & Sons Australia Ltd, 42 McDougall Street, Milton, Queensland 4064, Australia

John Wiley & Sons (Asia) Pte Ltd, 2 Clementi Loop #02-01, Jin Xing Distripark, Singapore 129809

John Wiley & Sons Canada Ltd, 6045 Freemont Blvd, Mississauga, ONT, L5R 4J3

Wiley also publishes its books in a variety of electronic formats. Some content that appears in print may not be available in
electronic books.

Library of Congress Cataloging-in-Publication Data
is available

British Library Cataloguing in Publication Data
A catalogue record for this book is available from the British Library

ISBN 978-0-470-51685-0 (PB)

Typeset in 11/13pt Baskerville by Sparks, Oxford – www.sparks.co.uk
Printed and bound in Italy by Printer Trento

This book is printed on acid-free paper.

Contents

Contents

Contents

Contents

Foreword by Giuliano Digilio
Head of Technical Services, Electrical Contractors' Association (ECA)

The IEE Wiring Regulations and more lately BS 7671 have always been important for electrical contractors and for installation designers, and they are a key factor in the implementation of electrical safety within the UK and indeed overseas. The IEE Wiring Regulations go back to the end of the 19th century, almost to the time of the very first electrical installation within the UK.

The ECA is fully committed to the development of standards for the national BS 7671 committee as well as corresponding work in both the European Committee for Electrotechnical Standardisation (CENELEC) and the International Electrotechnical Commission (IEC). This includes a considerable amount of work in the preparation for BS7671: 2008.

I am pleased that you have purchased the ECA *Guide to the Wiring Regulations* and I trust that this quality publication will aid to enhance the understanding and knowledge within the electrical industry for both electrical contractors and electrical designers.

Preface

This book discusses the requirements of BS 7671: 2008, also known as the IEE Wiring Regulations 17th Edition, published during January 2008.

The aim of the guide is to provide an explanation of the theory and reasons behind the Regulations, their meaning and the intent of their drafting. The book provides advice and guidance, demystifying the 'requirements' wherever possible. Practical and original solutions have been provided, which are often not found in other industry guidance.

The guide is a valuable resource for all users of BS 7671 including apprentices, electricians who perhaps want to 'dig a bit deeper' into the background of the Regulations, together with electrical technicians, installation engineers and design engineers. Most individuals who have any involvement with BS 7671 will find the book of considerable help and benefit in their everyday work.

To derive use and benefit from the book it is assumed that readers have knowledge of electrical installation engineering to a basic level. However, 'defined scope' installers and those at similar levels will also gain from working through the book thanks to its clear diagrams. Given these prerequisites, the book can be used as a learning text for the 17th Edition Wiring Regulations as long as readers have a copy of the Standard itself. Indeed, a copy of BS 7671: 2008 is required as a reference document when using this book, and readers should at least familiarize themselves with the terminology and definitions used in the basic Standard.

Guide to the Wiring Regulations is intended to be read on a chapter-by-chapter basis by those involved at the level of designing and constructing installations. This is something that is not easy to achieve with books on this subject as accessing the basic Standard itself can be quite daunting and heavy going.

A particular emphasis or expansion has been made to those subjects that are often confused by readers of BS 7671. In this respect the text does not wander off to discuss ancillary subjects; the text stays focused on providing an understanding of those concepts demanded of BS 7671 so that design and installation decisions can be made by readers.

The book's coverage is comprehensive, and all Parts of the Regulations have been addressed within the topic lead chapters. Design aspects have been included as they are integral to installations. Often, individuals or organizations consider themselves to be either pure designers or pure installers. However, even by the act of an 'installer' in selecting equipment that was unspecified by the designer, e.g. selecting cables or other equipment, an element of design is being carried out. The same concept is true of domestic installers who select 'standard designs' but perhaps feel that they do not design. These individuals are considered to be designers even where the design is not calculated for each installation. The adoption of a 'standard design' or a 'standard cable size' by the installer is in fact design by the installer.

The book is arranged into topic lead chapters, at the heart of which are Chapters C (Circuitry) and D (Selection and Erection of Equipment). Although the titles of these chapters seem simple enough, they are comprehensive and encompass about 70% of the Regulations.

Most requirements of the Regulations have been condensed and summarized using tables aided by clear, simple diagrams. Some tables seem quite long but they are still very condensed compared with the regulations that they summarize. As an example, the new Section 559 in BS 7671 includes 44 regulations, but these are summarized in a 15-row table. The nature of the Regulations is that they must state all facts. However, the repetition of the most basic information in the guide was not considered beneficial; for example, where the regulation is written in the following style:

'cables shall be large enough for the anticipated current'

This type of regulation is either not expanded upon in the guide or it is explained as follows:

'cables shall be $6\,\mathrm{mm}^2$ minimum'.

The book includes five printed appendices and further appendices are available as downloads from the companion website. Appendices that have been included on the Companion Website were either considered to be non-essential for most readers, or were items that may be subject to change at a future date. The Companion Website can be found at: http://www.wiley.com/go/eca_wiringregulations

Although more experienced readers may wish to jump to Chapter C, the introductory Chapters A and B are worth spending some time on. Within these chapters, the legal standing of BS 7671: 2008 is discussed together with its relationship with key UK law in the area of electrical installations. The general requirements of BS 7671: 2008 are also summarized within these chapters.

Acknowledgements

I would like to thank my wife Julie and my children for their patience, particularly throughout 2007, when much of the drafting of this book took place.

I also give particular thanks to Paul Cook, former staff member of IET, for his assistance with the Circuitry and Earthing and Bonding chapters, and to Leon Markwell, also former IET staff member, for his help with the Special Locations chapter.

I also thank James O'Neil, Director of Engineering of NG Bailey Limited, and Phil MacDonald, Principal Project Electrical Design Engineer of Shepherd Engineering Services, for acting as general readers, and Ken Morton, HM Principal Electrical Inspector, Health & Safety Executive, for his comments on Chapter B.

With thanks to David Thompson for the book design concept and for his creation of the illustrations.

Finally, I also wish to thank Simone Taylor, Nicky Skinner and their colleagues at the Wiley office, Chichester.

Darrell Locke, October 2007

BS 7671: 2008 – Introduction and Overview

A 1   Introduction to BS 7671: 2008

BS 7671: 2008 was published during January 2008 as a significant new Edition of this fundamental Standard.

Although the document is a British Standard, it is also known as (and jointly labelled as) the *IEE Wiring Regulations 17th Edition*; this is for copyright reasons. In spite of the fact that the IEE changed to the IET in 2006, the IET has maintained the brand of IEE, mainly for use in its Wiring Regulations documents and products. Indeed, the IEE logo appears on the front cover and the IET logo inside the front cover.

Throughout this book, BS 7671: 2008 is referred to as BS 7671: 2008, or variously as BS 7671, the Wiring Regulations, the Regulations, the 17th edition or the Standard, depending upon the particular context.

In essence, BS 7671: 2008 is virtually a European document. In fact, two parent documents as parts of the corresponding IEC standard have been used or adapted.

Guide to the Wiring Regulations: 17th Edition IEE Wiring Regulations (BS 7671: 2008)
Darrell Locke
© 2008 John Wiley & Sons Ltd

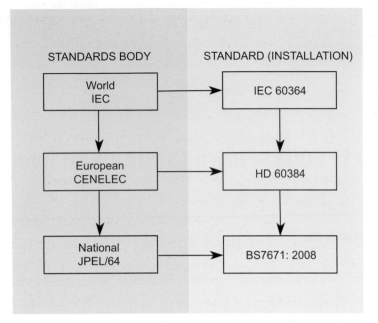

Figure A 1.1 Installation standards at world, European and national levels.

Both IEC and CENELEC have 'wiring regulation' standards or rules for electrical installations. The general structure of IEC, CENELEC and BS 7671 is illustrated in Figure A 1.1.

Many parts of the document originate in CENELEC in a 'harmonized document' (HD). The parent document is known as HD 60384 and comprises virtually all parts of the installation standard.

Within BS 7671: 2008 there are now only a few regulations that are truly 'UK only', although some of the CENELEC parts of HD 60384 have been modified, cut or expanded for BS 7671. Some of the appendices of BS 7671 are home-grown.

The Wiring Regulations committee has also used certain parts of the corresponding IEC document (IEC 60364) modified or virtually unmodified.

A list of the parts of the corresponding CENELEC parts of HD 60384 used in BS 7671: 2008 is shown in Table A 1.1.

Table A 1.1 Corresponding parts of CENELEC HD 60384 used in BS 7671: 2008.

CENELEC part	Issue date	Title	BS 7671 reference
prHD 60364–1	2007	Fundamental principles, assessment of general characteristics, definitions	Part 1, Part 2 (in part), Part 3
HD 384.4.41 S2/A1	2002	Protection against electric shock	Chapter 41
HD 384.4.42 S1 A2	1994	Protection against thermal effects	Chapter 42
HD 384.4.482 S1	1997	Protection against fire where particular risks or danger exist	Chapter 42
HD 384.4.43 S2	2001	Protection against overcurrent	Chapter 43
HD 384.4.473 A1	1980	Application of measures for protection against overcurrent	Chapter 43
HD 384.4.443 S1	2000	Protection against overvoltages	Section 443
prHD 60364–5-51	2003	Selection and erection of equipment – common rules	Chapter 51
HD 384.4.43 S2	2001	Protection against overcurrent	Chapter 53
prHD 60364–5-54	2004	Earthing arrangements, protective conductors and protective bonding conductors	Chapter 54
HD 384.7.714 S1	2000	Outdoor lighting installations	Section 559
HD 60364–7-715	2005	Extra-low voltage lighting installations	Section 559
HD 384.6.61 S2	2003	Initial verification	Part 6, Appendix 14
HD 60364–7-701	2007	Locations containing a bath or shower	Section 701
HD 384.7.702 S2	2002	Swimming pools and other basins	Section 702
HD 384.7.703	2005	Rooms and cabins containing sauna heaters	Section 703
HD 60364–7-704	2007	Construction and demolition site installations	Section 704
HD 60364–7-705	2007	Agricultural and horticultural premises	Section 705
HD 60364–7-706	2007	Conducting locations with restricted movement	Section 706
HD 384.7.708	2005	Caravan parks, camping parks and similar locations	Section 708
prHD 60364–7-709	2007	Marinas and similar locations	Section 709
HD 384.7.711	2003	Exhibitions, shows and stands	Section 711
HD 60364–7-712	2005	Solar photovoltaic (PV) power supply systems	Section 712
HD 60364–7-715	2005	Extra-low voltage lighting installations	Section 559
HD 60364–7-717	2004	Mobile or transportable units	Section 717
prHD 60364–7-721	200X	Electrical installations in caravans and motor caravans	Section 721
prHD 60364–7-740	2006	Temporary electrical installations for structures, amusement devices and booths at fairgrounds, amusement parks and circuses	Section 740

A 2 Plan and layout of BS 7671: 2008

Most users will not need to concern themselves with the correct terminology for groups of regulations and chapters etc., but an explanation of this has been added for completeness.

Let us look at a single Regulation 411.3.2.1 and provide a diagram of the structure. Taking the first three digits, these relate as follows:

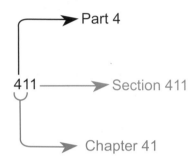

The remaining numbers make up the group, sub-set and regulation, but really only the group is of any significance:

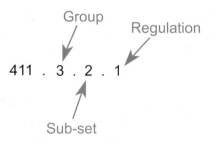

There are seven parts to BS 7671: 2008 as follows:

Part	Contents
1	Scope, Object and Fundamental Principles
2	Definitions
3	Assessment of General Characteristics
4	Protection for Safety
5	Selection and Erection of Equipment
6	Inspection and Testing
7	Special Installation or Locations

A3 Overview of major changes

There is not much of the document which remains unchanged compared with the 16th Edition; many changes were due to formal incorporation of CENELEC drafts required to achieve harmonization.

This section gives an overview of technical changes that will manifest a change in practice or will be something that you should be aware of. As stated in the preface the subject of BS 7671 can be heavy going and this part of the book has been kept as short as possible. Readers may wish to skip this part of the book and start with the two key chapters, C and D.

The following overview notes have been included, and are listed in what is considered to be an 'order of significance'.

Chapter 41 Protection against electric shock

Revision of Chapter 41 is probably the most significant revision made for the 17th Edition.

The whole structure of the chapter has been modified. The familiar terms used in the 16th Edition of 'direct contact' and 'indirect contact' have been replaced with 'basic protection' and 'fault protection' respectively. This terminology change by itself had ramifications in many other parts of the Regulations and these brought about logistical modifications. The various measures are termed 'protective measures'.

The structure of Chapter 41 was accordingly modified. Basic protection (insulation and enclosures) was considered something that designers and installers did not actually 'consider' and was shunted towards the rear of the chapter. The extremely rare measures of 'placing out of reach', 'obstacles', 'non-conducting location', 'earth-free local equipotential bonding' and 'electrical separation' were shunted further to the rear of the chapter. Thus, the main reading in the front end of Chapter 41 is about automatic disconnection.

There have been changes to disconnection times. There are no 'mixed' disconnection times and disconnection times have been introduced for TT installations. As protection in TT installations will virtually always require an RCD, the reduced disconnection times in the 17th Edition are easily achieved (0.2 s for final circuits).

A very significant new Regulation (411.3.3) requires a 30 mA RCD for socket-outlet circuits that are intended for use by ordinary persons. With a few exceptions, this

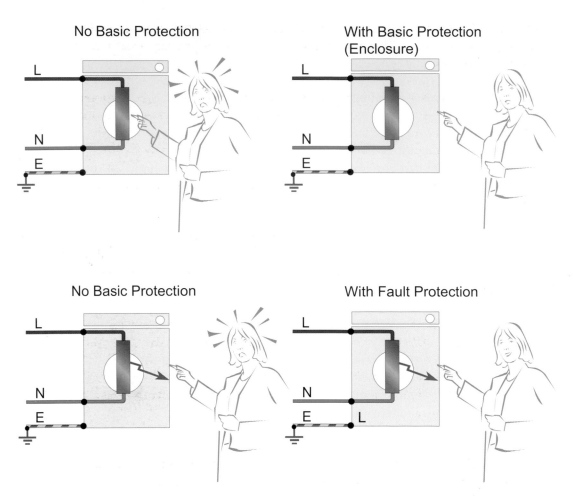

No Basic Protection

With Basic Protection
(Enclosure)

No Basic Protection

With Fault Protection

Figure A 3.1 Basic protection and fault protection.

means all domestic installations. Commercial installations will generally remain exempt, as in most such situations individuals will have received instruction.

Guidance on the structure, disconnection times and the use of RCDs is given in Chapter C of this book.

Bathrooms 701
The 17th Edition goes the extra mile on harmonization with CENELEC for bathroom installations.

The 16th Edition introduced the concept of 'Zones' to the regulations for bathrooms but fell short of harmonization with Europe in one key area: socket outlets in bathrooms.

Section 701 now aligns with the European ethos; there is no Zone 3. Thus, outside Zone 2, which is 600 mm from the bath or shower, only the 'general rules' of the regulations apply and any equipment is allowed, although socket outlets have a special distance specified and must be at least 3 m from the boundary of Zone 1.

All bathroom circuits now require a 30 mA RCD and a UK modification negates the need for supplementary bonding.

Tables and methods of cable current-carrying capacity (Appendix 4 of BS 7671)

The whole of the front end of this appendix has been modified for the 17th Edition. The modifications include the following:

- Overhaul of the installation methods and reference installation methods.
- New installation methods for cables in domestic style insulated cavity floors and lofts.
- 'Rating' factors for cables buried in the ground.
- Extensive additional rating factors for cables in free air (called correction factors in the 16th Edition).

Swimming pools (702)

For the 17th Edition, the scope of section 702 now includes the basins of fountains and areas in natural waters including the sea and lakes, where they are specifically designated as swimming areas.

Lighting and luminaires

A completely new section for the 17th Edition is section 559 'Luminaires and Lighting Installations', which totals six pages of text and some 44 new regulations.

The new section deals with interior and exterior lighting installations and also applies to highway power supplies and street furniture.

The section specifies such regulations as: 'luminaire through wiring can only be used where the luminaire is specifically designed for this', 'heat specification of terminal wiring', and others of a similar nature.

Inspection and testing

Under this heading a new requirement is that the insulation resistance of conductors is tested to the cpc with the cpc connected to the earthing arrangement.

New appendix with current-carrying capacity of busbars

A new appendix has been added giving information on current-carrying capacity and voltage drop of busbars and powertrack.

Chapter 56 Safety services

This chapter has been modified and specifies 'break times' for standby systems. It sets regulations for such subjects as circuitry under fault conditions, parallel operation, and specifies the life of certain critical back-up batteries.

High earth leakage currents

Correctly termed 'high protective conductor currents'. The former section 607 has been incorporated into Chapter 54 with some limited removal of ambiguous regulations.

High voltage to low voltage faults

A new section for the 17th Edition, but this not particularly significant for installers or designers. The section is only relevant for 'private' HV-LV substations, and even then the corresponding HV standards will need to be followed. Read Chapter D for a fuller explanation.

Voltage drop

Whilst in essence the basic requirements of the regulations on voltage drop have not changed, a new appendix suggests maximum voltage drops for both utility and private supplies. These voltage drops are separated into suggested limits for lighting and other circuits.

Atmospheric and switching overvoltages

There are a few pages of regulations on this subject, but there is not much of significance unless you have overhead distribution cables within your installation.

Surge protective devices

Although not required, there are regulations for installing surge protective devices (SPDs).

Insulation monitoring devices (IMDs) and residual current monitors (RCMs)

Similarly, although optional, there are regulations for installing these devices.

RCMs in particular are becoming more widely specified, and there is guidance on this subject provided in Chapter D of this book.

Caravan and camping parks (708)

The main modification for the 17th Edition is that pitch socket-outlets are to be individually protected by a 30 mA RCD.

New special installations or locations

The following Special Installations sections are new to the 17th Edition:

- 709 Marinas
- 711 Exhibitions, shows and stands
- 712 Solar photovoltaic (PV) power supply systems
- 717 Mobile or transportable units
- 721 Electrical installations in caravans and motor caravans
- 740 Temporary electrical installations for structures, amusements and booths at fairgrounds
- 753 Floor and ceiling heating systems.

Legal Relationship and General Requirements of BS 7671: 2008

Introduction

It is important to recognize that, for electrical designers and installers, there are legal responsibilities that must be both known and implemented whilst carrying out electrical installation or electrical design work. This chapter provides information and guidance on key UK legislation relevant to electrical installations. It also provides guidance on some contractual obligations relating to designs and installations.

The chapter is neither a full legal guide nor a full contractual guide to requirements but provides a short overview.

The chapter finishes with notes on the assessment of general characteristics (i.e. Part 3 of BS 7671: 2008).

B 1 Legal requirements and relationship

B 1.1 Key UK legislation

Legislation can be in the form of an Act of Parliament (e.g. The Health & Safety at Work etc. Act 1974) or a Statutory Instrument (e.g. The Electricity at Work Regulations 1989). Acts are primary legislation and Statutory Instruments are secondary legislation made under a specific Act – in the case of the Electricity at Work Regulations, these were made under the Health & Safety at Work Act. Failing to comply with requirements of an Act of Parliament or a Statutory Instrument is a breach of criminal law and may result in a prosecution.

The following legislation is considered key and relevant to the electrical technical aspects of electrical designs and electrical installations:

- The Electricity at Work Regulations 1989
- The Electricity Safety, Quality and Continuity Regulations 2000 (as amended 2006)
- The Electricity Act 1984
- The Building Acts 1984 & 2000 (These apply to England and Wales only and implement the Building Regulations for England and Wales including Approved Document Part P (Electrical Safety – Dwellings).
- The Building (Scotland) Act 2003 (This applies in Scotland only and implements the Building (Scotland) Regulations 2004.
- The Electromagnetic Compatibility Regulations 2006
- Tort

B 1.2   The Electricity at Work Regulations 1989 (EWR 1989)

The EWR 1989 is one of the most important pieces of legislation that an electrical designer or electrical contractor must be familiar with. You should know the content of this document as well as know of its existence.

The EWR 1989 covers the safety of people, including employees, involved in all aspects of electrical and electronic systems in the UK. This includes self-employed electricians working in domestic installations; all 'electrical personnel' in commercial installations and construction sites; and for commercial installations the end users. It also includes any person undertaking any work activity on or near electrical equipment.

All electrical equipment and systems are encompassed by the legislation, from a battery to the national super grid at 400 kV. The legislation covers design, operation, isolation, maintenance, workspace and lighting equipment. There are Regulations on precautions for working on equipment made dead and on work on, or near, live conductors. There are also requirements for persons undertaking work to be competent to prevent danger and injury.

Compliance with EWR 1989 is therefore a fundamental requirement for any organization, and it is recommended that organizations have in place a system of training to ensure compliance with the Regulations. Guidance on EWR 1989 is available from the Health & Safety Executive (publication HSR 25 – *Memorandum of guidance on the Electricity at Work Regulations 1989*). It is recommended that organizations purchase this and implement the guidance provided; it contains the text of the Act as well as HSE guidance on how to achieve the statutory requirements. The document details and availability are as follows:

> *HSE's 'Memorandum of Guidance on the Electricity at Work Regulations 1989 – HSR 25'*
>
> *ISBN 9780717662289*
>
> *(50 pages, available from HSE Books and is referenced HSR25)*

As well as having a copy of HSR 25, organizations must provide adequate training for all staff that work on or near electrical installations.

To supplement this, a dead working policy should ideally be formalized together with a live working policy for those contractors that carry out live work.

In respect of the 'making dead or live' working aspects of EWR 1989, the following document is also very useful, if not essential.

> *HSG 85 Electricity at Work – Safe Working Practices*
>
> *ISBN 0717621642*

This document expands upon detail of policy and procedures for safe working practices for people who work on or near electrical equipment. It includes guidance on the following:

- assessing safe working practices;
- deciding to work dead or live;
- actions common to both dead and live working;
- working dead; and
- working live.

B 1.3   The Electricity Safety, Quality and Continuity Regulations 2002 (as amended)

These statutory Regulations are primarily intended for distribution network operators (DNOs), setting statutory limits for voltage and frequency.

The Regulations state that PME supplies cannot be used to supply installations supplying caravans or boats. Also, DNOs can take the option not to provide an earth to installations that they feel are inappropriate. This will possibly be the case on some farms, building sites and petrol filling stations.

For all installations, DNOs will have to take a view on the safety of an installation and will use BS 7671: 2008 for this. If the DNO feels that an installation is unsafe, they can refuse to provide a supply or, if connected, disconnect the supply.

B 1.4 The Electricity Act 1984 (as amended)

This Act is primarily aimed at distribution network operators and (more recently) meter operators. However, there is a relevant point for installation designers and contractors to note: the Act gives the DNO or meter operator the right to position intakes where they see fit and where they feel appropriate for a given installation.

B 1.5 The Building Act 1984, The Building Regulations and Part P

The Building Act 1984 refers only to England and Wales. This book does not cover all the technical requirements relating to the Building Regulations. There are numerous guides and books on this subject including one produced by the ECA and the NICEIC.

However, Part P of the Building Regulations, on the subject of electrical safety within dwellings, is summarized in this section.

Legal standing of Part P
The Building Regulations are made under the main Act of Parliament, the Building Act 1984. The Building Act is the primary legislation and itself refers to the Building Regulations 2000 with its various Parts on structure, means of escape, spread of fire, ventilation, heat loss and, of course, electrical safety.

The Building Regulations 2000 are statutory and a breach of the Regulations in itself is an offence under criminal law.

As mentioned earlier, statutory instruments such as the Building Regulations must be complied with, otherwise a breach may result in a prosecution.

For guidance, and to specify a recognized way of complying with the individual parts of the Building Regulations, the CLG (Communities for Local Government) produces 'Approved Documents' on each part of the Building Regulations. It is important to recognize that the Approved Documents themselves are not statutory. This is demonstrated in Figure B 1.1.

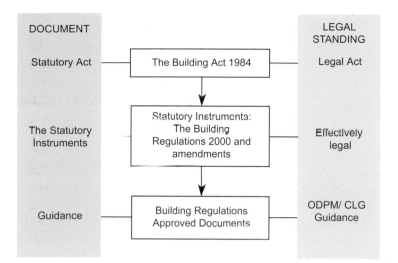

Figure B 1.1 Relationship of Building Act, Statutory Instruments and Approved Documents.

The wording of the 'Statutory Instrument' and hence the legal requirement of Part P is given in Table B 1.1.

Table B 1.1 Statutory Instrument relating to Part P of the Building Regulations.

Requirement	Limits on application
PART P ELECTRICAL SAFETY	
Design, installation, inspection and testing	
P1 Reasonable provision shall be made in the design, installation, inspection and testing of electrical installations in order to protect persons from fire or injury. **Provision of information** P2 Sufficient information shall be provided so that persons wishing to operate, maintain or alter an electrical installation can do so with reasonable safety.	The requirements of this Part apply only to electrical installations that are intended to operate at low or extra-low voltage and are in a dwelling; in the common parts of a building serving one or more dwellings, but excluding power supplies to lifts; in a building that receives its electricity from a source located within or shared with a dwelling; and in a garden or in or on land associated with a building where the electricity is from a source located within or shared with a dwelling.

B 1.6 The Electromagnetic Compatibility Regulations 2005 (EMC)

Information on EMC and duties under the EMC directive and BS 7671: 2008 are provided in Chapter D section D 4.

B 1.7 Tort and negligence

This section is included, not to scare, but to provide information on this 'common law' duty.

Tort

The English legal system recognizes two forms of 'wrongs'. The first is what we call 'criminal' and may be punished by a fine, imprisonment or both. Commonly we think of murder, but this act stemmed from an infringement of society's moral code, i.e. it was morally wrong to deprive someone of their life. These moral codes were identified and legislated for, initially by the overlords, then monarchs, and ultimately by Parliament. The Health & Safety at Work etc. Act 1974 is Parliament-enacted law and, as such, creates a criminal obligation upon any transgressor.

The second is called 'civil' or 'common' law, colloquially known as 'judge-made law' because the rules and principles have been created in the courts of the land, enshrined by what is termed 'the law of precedent'; that is, unless overturned by a superior court, the ruling establishes the law and binds judges in any subsequent cases. The civil law is concerned with providing restitution of rights, obligations or finances in the event of some form of dispute, termed a breach.

Civil law governs both the circumstances where there is an intention to form a relationship, by creation of a legally binding agreement – we call this the law of contract – and where a relationship may exist but where no contract is present, which we call the law of torts. Tort may thus be considered liability where there is no contract.

Torts include negligence, nuisance, defamation and trespass, to name but a few. It is possible to owe a duty in both tort and criminal law. However, if an action is successfully pursued in a criminal court, the 'beyond reasonable doubt' burden of proof being considerably higher than the civil determination of 'viewed against the balance of probabilities', means that the civil liability is taken as having been established. It is the function of the criminal compensation board to establish the level of civil damages due to a 'common' infringement of rights, having established a criminal liability.

In tort there is a need to establish a relationship. The landmark case is Donoghue v Stevenson (1932), wherein a friend of Donoghue purchased for her a bottle of ginger beer, found to contain a partially decomposed snail. Donoghue then successfully established that the manufacturer owed her a 'duty of care' under the tort of negligence, and hence was in a relationship, even though 'she' had no direct contractual relationship with the manufacturer of her own. This 'neighbour' principle is important as it establishes whether a duty of care is owed or not. The question follows: Who is my neighbour? Well, the answer is anyone who it is foreseeable to be likely to be affected by your actions.

You can see that liability in tort is therefore very wide, and the rules governing its implementation are extremely complex. Seventy years on, the courts are still grappling with the principles and extent of this law.

To some extent this is why 'collateral warranties' are called for, because rather than rely on the tort of negligence, parties in a collateral warranty agreement can instead sue for a breach of contract. The level of damages may be similar or higher and it is easier to prove a breach under contract law. We ignore 'pure economic loss' and the newly enacted Rights of Third Parties Act.

Negligence

If you negligently design a system or provide a service, and as a result it causes death or personal injury, or causes damage to other property, then you can be held liable for these losses under the tort of negligence. Making a mistake, or getting something wrong, is not being negligent. Being negligent is where you are found to have performed at a level less than would have been expected by a 'reasonable man' whilst undertaking a task, where you had held yourself out as being competent to undertake that task. Thus, if you hold yourself out as being competent to design a lighting system, offer advice concerning that system, and others rely on that advice and install what is subsequently found to be deficient, then irrespective of payment, you may still be held financially liable. It is for this reason that services designers and contractors are strongly advised to insure themselves with professional indemnity insurance.

B 2   The role of Standards

Definition of Standards

Standards, including international, European and British Standards, are documents to bring about simplification, interchangeability, terminology, methodology, specification or codified practice.

A Standard is defined as:

> 'A document established by consumers and approved by a recognized body that provides for common and repeated use, rules, guidelines or characterization for activities or their results, aimed at the achievement of the optimum degree in a given context.'
>
> (This was taken from IEC Guide 2, 1996)

Standards are (or should be) written by industry by consensus where consensus is defined as:

> 'General agreement, characterized by the absence of sustained opposition or substantial opposition by any important part of the concerned interests and by a process that involves seeking to take into account the views of all parties concerned and to reconcile any conflicting arguments.'

Legal standing of Standards

Standards are described or implicated by statute. Standards are voluntary codes of rules, and are not law nor are they legally enforceable. Indeed, individuals may take a view to ignore a particular standard. However, some standards are boosted to an elevated status by being referred to either directly or indirectly in statutes. Depending upon the wording, this can make the standards themselves have a quasi-legal status. Again, though, there is a caveat. A good way to explain this further is to look at how BS 7671 is referred to in some legal documents.

BS 7671 and the Electricity at Work Regulations 1989 (EWR)

It is important to recognize that the wording of the EWR makes no mention of BS 7671. The HSE's Memorandum of Guidance (HSR 25) states that:

> 'BS7671 is a code of practice which is widely recognized and accepted in the UK and compliance with it is likely to achieve compliance with relevant aspects of the 1989 (EWR) Regulations.'

BS 7671: 2001 in Part P of the Building Regulations 2000

In a similar fashion, BS 7671 is not mentioned in the primary legislation, which simply states that 'the installation shall be designed and installed in order to protect persons from fire or injury'.

It is Approved Document P (which itself is guidance) that mentions BS 7671.

This states 'In the Secretary of State's view, the (Part P) requirements will be met by adherence to the 'Fundamental Principles' for achieving safety given in BS

7671: 2001 Chapter 13'. This will need updating to read correctly for BS 7671: 2008.

Standards implied or prescribed by contract

Standards are often prescribed by a contract on a definite item (stated) or by a general contract term similar to 'shall comply with all relevant codes and standards'.

Assuming the standard is relevant or if it is listed, then compliance with the Standard becomes binding under the UK law of contract.

B 3 Part 3 of BS 7671: 2008 – assessment of general characteristics

Part 3 of BS 7671, totalling only four pages, sets requirements for an overall assessment of an electrical installation. It is intended that the requirements of Part 3 be considered *prior* to the design of an installation in compliance with other Parts of BS 7671. This works for some of the regulations in Part 3, but some are really repetitive of the general requirements given in Parts 4 or 5.

The requirements are summarized in Table B 3.1 in just five paragraphs. The regulation numbers have been omitted here for clarity and due to the fact that the requirements are so general.

Table B 3.1 Requirements of Part 3 of BS 7671: 2008.

Requirement of Regulations	Notes and advice
The installation shall be assessed for purpose, external influence, compatibility, maintainability, continuity of service and recognized safety services	
The characteristics of voltage, current, frequency, prospective fault current, earth fault loop impedance (ELI), maximum demand and protective device at the origin shall be determined	This can be done by inspection, by enquiry, measurement, calculation and applies to all sources of supply. Safety supplies shall be assessed separately and the requirements of these are in Chapter 56 of BS 7671; see Chapter D
Installations shall be suitably divided up to avoid danger, minimize inconvenience in the event of a fault, reduce the possibility of unwanted tripping of RCDs, mitigate the effects of electromagnetic interference, and ensure effective isolation Continuity of supply for the intended use and life of the installation shall be considered	These requirements are discussed with recommendations made in Chapter C
Final circuits shall be connected to separate protective devices at distribution boards	
Compatibility and EMC shall be considered	See Chapters C and D

Circuitry and Related Parts of BS 7671: 2008

C 1 Introduction

This chapter provides information and guidance on circuitry aspects of BS 7671: 2008 and includes a significant amount of regulations associated with circuits. For example, 'protection against electric shock' aspects are included. Like other chapters, the structure is topic led and can be read in page order. The chapter guides you through what you need to design and install circuits to BS 7671: 2008 and applies to significantly large installations. It does not cater for very large or complex installations with, for example, interconnecting busbars, and such complexity is outside the scope of this book.

There is a certain amount of overlap with Chapter D, and these two chapters should both be read prior to undertaking design or installation.

Lastly, in this chapter, unlike other chapters, there are not numerous references to individual regulation numbers. This is due to the fact that most of the circuitry aspects are covered by relatively few regulations in BS 7671: 2008, the importance of which cannot be overstated. Extensive background knowledge and understanding is required to comply with these regulations and this chapter guides readers through all relevant aspects needed.

C 2 Design procedure overview

The procedure of carrying out an electrical system design of an installation can be quite involved and often a number of drafts and subsequent adjustments are necessary. The following flow diagram shows the logical order of steps in the design process.

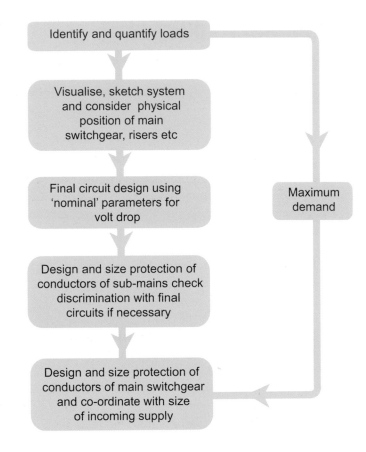

By studying this flow chart it should be obvious that a certain number of iterations with adjustments will be required, as a system is rarely designed without a certain amount of to and fro.

The contents of this chapter will need full consideration and, when carrying out a design in practice, adjustments will be necessary at various stages of the process.

Discrimination between all upstream and downstream protective devices may be required for convenience or continuity of supply to essential equipment, but this may make the electrical system over-designed (much too large for its designed use) and thus carry a cost burden (see C 6.1).

To provide for a cost-effective and efficient design it helps if the main incoming supply point is close to the load centre of the installation, and hence discussions with the electricity distributor should be started at an early stage.

It is not essential that the main distribution board(s) are positioned close to the intake point, and their position has an effect on voltage drop on the whole installation including the submain cables.

This point of 'positioning' is also true of final circuit distribution boards which need to be carefully considered in terms of voltage drop in large installations with highly loaded final circuits. The concept of how to achieve this will become clearer when this chapter has been read.

 ## C 3   Load assessment

C 3.1   Principles and definitions

The subject of load assessment often comes down to experience, and there is no substitute for this.

Many installations have major identifiable loads. In commercial premises these usually include air-conditioning with chillers, heater banks, compressors and motors of all types as well as lifts, lighting and user 'final' equipment loads usually served by socket outlets.

A 'newer' load is the electrical supply to 'data storage' facilities (data centres). Although beyond the scope of this book, data centres require vast amounts of power, but between a large purpose-built data centre and an installation with a few PCs there are installations with small and medium data storage or server rooms. These have notable electrical power and cooling loads, and these loads should be considered.

Interestingly on this subject, whilst traditional loads such as lighting have become more efficient, overall supply demand has increased due to computing and data processing loads.

So, how is an installation's electrical load quantified and estimated?

Firstly, it is important to clarify the terms used, as some of these are not defined in BS 7671.

Connected load

Connected load (or total connected load) is taken to be the sum of all loads in the installation.

Care is needed in specifying this load; diversity (See section C3.3) cannot be used but duty cycle (cyclic load) considerations may need to be included.

Duty cycle

For a device or piece of equipment used intermittently, this is the cycle of starting, operating and stopping. Also included is the time interval that elapses during such a cycle.

Alternatively expressed, for a device or piece of equipment used intermittently it is the ratio of its operating time to its rest time, or to total time.

Crest factor

In a periodically varying function (such as that of a.c.) this is the ratio of the peak amplitude to the RMS amplitude.

Some may know this definition as 'load factor', and is not the same as duty cycle. Both terms are further explained with the aid of an example. This example would be needed for cyclic loads (533.2.1) evaluation.

Consider an installation with two motors of the same type installed in different applications. One motor is used in a supply air fan, the other in a passenger lift application. Both motors have a 20 kW motor with a full load running current of 35 amps and a starting current of 175 A. The lift is in a busy, frequently visited building, particularly busy between 9.00 am and 9.30 am.

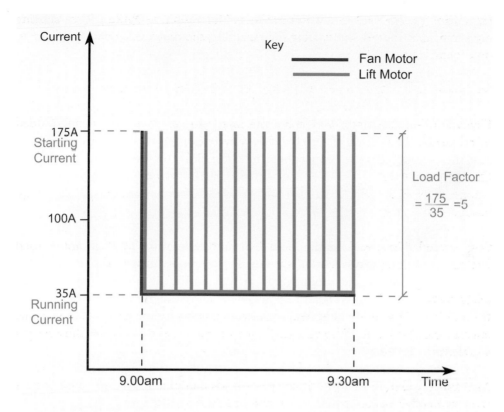

Figure C 3.1 Starting and running currents of lift motor and fan motor between 9.00 am and 9.30 am.

The duty cycle and crest factor for both motors is shown in Figure C 3.1.

Figure C 3.1 shows that both motors have the same crest factor but very different duty cycles. The fan's duty cycle can be taken as unity (1) and its starting current ignored in terms of heating effect on the supply switchgear and cables. The starting current of the lift motor, however, needs consideration, as there are 15 starting occasions in 30 min. If the starting time was 30 s, the duty cycle would be

$$\frac{0.5 \times 15}{30} = 0.25$$

indicating that the motor was in starting mode for 25% of its time. From this information, the heating effect of the duty cycle can be calculated using the fact that the heating effect is proportional to I^2Rt. Thus, the equivalent continuous current is

$$I_{\text{equivalent}} = \sqrt{i_1^2 R_1 t_1 + i_2^2 R_2 t_2} = \sqrt{35^2 \times 0.75 + 175^2 \times 0.25} = 93\,\text{A}$$

This result demonstrates the significance of duty cycle of the cyclic lift motor load compared with the steady load of the fan. The lift motor supply would need to be rated at 100 A, and the fan motor supply at 50 A (for the lift motor running current of 35 A). It should be noted that in practice the lift motor starting current will not be on for 30 s, this figure being exaggerated to emphasize the point.

Maximum demand

Maximum demand (or maximum power demand) is the highest rate at which power is consumed. Alternatively expressed, it is the highest average rate at which electrical power is consumed.

In calculating the maximum demand in an installation, diversity can be applied (311.1).

To apply these definitions to an installation still requires appropriate experience and usually a lack of such experience leads to an overdesigned electrical distribution system. The author has attended many installations with 500% or greater overcapacity with all the client's equipment connected and running.

C 3.2   Maximum demand assessment

Maximum demand is sometimes expressed as:

maximum demand (kVA) = connected load × diversity

Diversity is discussed in section C 3.3.

There are two methods for calculating maximum demand, as follows:

- A summation of individual connected loads with application of diversity factors.
- Comparison with table of 'norms' for similar installations.

In practice, a combination of these two methods is often utilized. For example, major plant loads are calculated with duty cycle and diversity applied and this is added to the 'norm' watts per square metre for general areas of the installation.

For assessment by adding individual loads there are nearly always differences between the rated power expressed by the manufacturer, and the actual currents drawn. This can be true despite checking catalogues and data sheets as well as rating plates and this contributes to overdesign. More often than not, actual connected loads are 50–70% of the quoted value.

The watts per square metre method can be used to produce an overall maximum demand estimate, alongside information on known loads, or it can be used solely to produce an estimate.

The watts per square metre method involves comparing the installation type and size with a watts per square metre 'normal' table. Good in theory, but where does the table come from? There have been limited central studies of this type of 'norm' in the UK, although consulting design firms and larger contractors collect their own data. CIBSE (the Chartered Institution of Building Services Engineers) does publish a certain amount of such data in its journal, but this information is connected load and not actual running load. BSRIA (www.bsria.co.uk) also publishes some guidance on this subject.

Table C 3.1 provides an assessment of load and would equate to the maximum demand estimate.

The table can be used for generic designs while noting that it is an 'average norm' table, and if you have significant loads not found in the average office the table will be inaccurate. Medium-sized data centres are excluded from this table.

Table C 3.1 Typical load assessment for commercial offices.

Office size	Load no air-conditioning (W/m²)	Load with air-conditioning (W/m²)
Small office Up to 2000 m²	70	120
Medium office 2000 m² to 10 000 m²	60	110
Large office Over 10 000 m²	55	100

C 3.3 Diversity

311.1

Diversity should be taken into account when assessing the maximum demand of an installation (311.1).

Diversity is the engineering principle that in any given installation, some of the connected loads will not be running at the same time instant as other loads. This principle can be further broken down into two types of load as follows:

A Loads that, due to the law of averages, will not be on at the same time.
B Loads that, due to fact, will not be on at the same time.

Examples of type A include instantaneous electric showers in a multiple block of flats, lift supplies in general and motors for building services. Examples of type B include electrical heating loads and electrical cooling loads; obviously, while it is possible to run both together, the fact is that they do not. There are many examples of both types of load.

In attempting to make an assessment of diversity, there is no substitute for knowledge and experience. The extent of knowledge and experience needed must match the type of installation being assessed.

It should be recognized that diversity can be applied in a number of ways as follows:

● for items on a final circuit (except socket outlets);
● between similar final circuits, i.e. assume one circuit is 100%, the other 0% or $x\%$;
● between sub-distribution boards or submain cables; and
● at each main distribution board.

Table C 3.2 indicates some suggested diversity factors for average circumstances and can only be used by those with suitable experience and knowledge of the type of installation being assessed.

It should also be noted that many engineers, technicians and electricians are inclined to significantly overestimate loadings – perhaps to play safe – and this leads to an overdesigned electrical system. It is more skilful to produce an ample design with capacity built-in, but which is not grossly overdesigned. An interesting point to note here is the supply utility's figures for domestic supplies; they use a figure for domestic maximum demand of about 2 kW as an average consumer load. This figure says much about diversity when applied correctly.

It should be noted that Table C 3.2 may differ from other published data on the subject; it is felt that Table C 3.2 is realistic, subject to the constraints given above on experience.

Table C 3.2 Some suggested diversity factors.

Item	Diversity factor	Notes
Lighting in small office and similar, up to 2000 m²	0.7	0.6 with daylight control
Lighting in medium office and similar, 2000 m² to 10 000 m²	0.8	0.7 with daylight control
Lighting in large office and similar, over 10 000 m²	0.85	0.7 with daylight control
Retail store lighting	0.9	
Space heating in small office and similar, up to 2000 m²	0.8	Capacity of system and 24 hour cycle to be considered for thermal capacity; adjust as necessary
Space heating in medium office and similar, 2000 m² to 10 000 m²	0.7	Capacity of system and 24 hour cycle to be considered for thermal capacity; adjust as necessary
Space heating in large office and similar, over 10 000 m²	0.6	Capacity of system and 24 hour cycle to be considered for thermal capacity; adjust as necessary
Socket outlets – all commercial general purpose office, all sizes	Use W/m²	More appropriate to use the overall table figures in Table C 3.1

C 4 Circuit design

C 4.1 Introduction

This section explains how to carry out cable sizing manually. Modern cable sizing software programs can be quite sophisticated and, for most projects, save considerable time. However, as engineers you must know if the inputs you are making, as well as the outputs that you are receiving, are correct.

In this section, the design procedure common to all circuits is considered. With reference to BS 7671, there are four separate subject areas that will determine the cable size as follows:

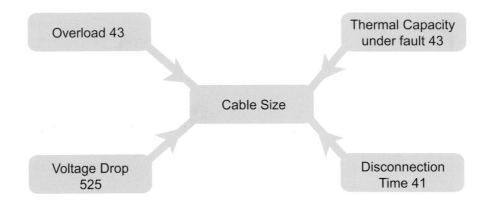

These factors are considered and discussed in this chapter. Readers should note that ultimately only one of these factors will, at any point and for a particular circumstance, determine the cable size. Experience and some rules of thumb given in this chapter may, for example, lead you to carry out a voltage drop check in preference to one of the other sizing factors.

In order to visualize the cable sizing process, Figure C 4.1 provides a flow chart of the process showing the order of stages.

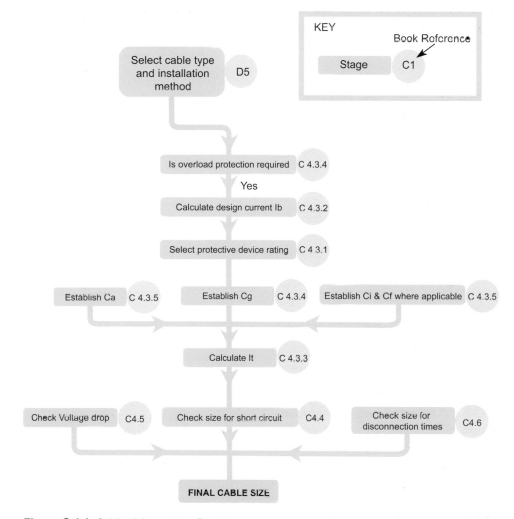

Figure C 4.1 Cable sizing stage diagram.

C 4.2   Protection against overcurrent in general

Protection against overcurrent as defined in BS 7671: 2008 includes overload and fault currents:

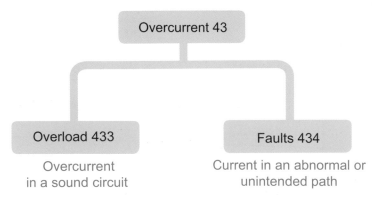

It should be noted that an earth fault current in terms of BS 7671: 2008 is simply known as a fault current; but this can cause confusion, and the term earth fault current is used in this book. The earth fault current requirements specified in Chapter 54 of BS 7671 are in essence the same as the short circuit requirements specified in Chapter 43, and aspects of protective conductor sizing are therefore included in this part of the book.

C 4.3   Overload protection

C 4.3.1 Fundamentals

There are circumstances where overload protection is not required – usually where the continuity of the supply is critical compared with the implications of not providing overload protection. These are discussed later.

For most circuits overload protection is required, and this protection should become second nature to installation engineers.

A basic requirement given in 433.1 is that circuits should be arranged so that small overloads for long durations are 'unlikely to occur'. What is meant by small overloads are those that will not be detected by the protective device. For MCBs this would be somewhere between 1 and 1.45 of the rated current of the device (see D 6.2).

Problems arising from not complying with this regulation are rare (cables are usually oversized) and only an error in obtaining an accurate current rating would cause difficulties. Another fundamental requirement of 433.1.1 is that:

$$I_b \leq I_n \leq I_z$$

where

I_b is the design current of the circuit,

I_n is the nominal current or current setting of the protective device,

I_z is the current-carrying capacity of the conductor in the particular installation conditions.

Pictorially represented this requirement is as follows:

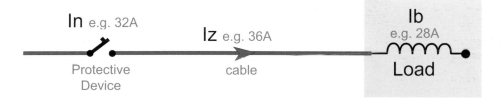

Devices will operate if the current exceeds the fusing or tripping current I_2 for a time greater than the conventional fusing or tripping time. Fusing, non-fusing and conventional times are given for common devices in Table C 4.1.

Table C 4.1 Fusing and non-fusing currents and conventional fusing times.

Device type	Rated current I_n (A)	Non-fusing or non-tripping current I_1 (A)	Fusing or tripping current I_2 (A)	Conventional fusing or tripping time (h)
MCBs to BS EN 60898	≤ 63 ≥ 63	1.13 In 1.13 In	1.45 I_n 1.45 I_n	1 2
BS 88 fuse	<16 $16 < I_n \leq 63$ $63 < I_n \leq 160$ $160 \leq I_n \leq 400$ $400 < I_n$	1.25 I_n for 1 h 1.25 I_n for 1 h 1.25 I_n for 2 h 1.25 I_n for 3 h 1.25 I_n for 4 h	1.6 I_n 1.6 I_n 1.6 I_n 1.6 I_n 1.6 I_n	1 1 2 3 4
BS 1361	$5 < I_n \leq 45$ $60 < I_n \leq 100$		1.5 I_n 1.5 I_n	4 4

Overcurrent devices will not operate if the overcurrent is of short duration (except for circuit breakers where the overcurrent exceeds the instantaneous operating current).

C 4.3.2 Design current I_b

In order to determine a correct individual cable size it is important to obtain or calculate an accurate circuit design current. This may not be critical for short or lightly loaded circuits, but becomes critical when sizing heavily loaded circuits with long cable runs of, say, over 75 m. At longer circuit lengths, voltage drop requirements can lead to cable sizes significantly larger than the base size. In order to minimize this in calculations, an accurate design current I_b can be key in ensuring a good design.

The basic formulae to apply to obtain a design current are as follows:

$$\text{Single-phase equipment, } I = \frac{kW}{V \times pf} \times 1000 \text{ (amps)},$$

$$\text{Three-phase, } I_1 = \frac{kW}{\sqrt{3} \ V_1 \times pf} \times 1000 \text{ (amps)},$$

where:

V is the nominal phase voltage to earth, also denoted by U_0

V_1 is the line voltage, also denoted by U

pf is the power factor

I_1 is the line current in a three-phase system.

C 4.3.3 Installed cable sizing

While carrying out cable sizing to BS 7671, or indeed for cable sizing in general, it helps to bear in mind the following principles:

- An insulated cable size is only limited by its type of insulation.
- For a given insulation, the rating depends upon both load current and the rate of heat dissipated by the cable to its immediate environment.

The basic sizing principle is as follows:

First select an appropriate protective rating, larger than the design current (step 1).

Next the initial, tabulated cable size (I_t) is obtained by using the protective device rating and correcting for ambient temperature, grouping factor and, where applicable, correction factors for thermal insulation and rewirable fuses (step 2).

$$I_t = \frac{I_n}{C_a \, C_g \, C_i \, C_c}$$
(Equation 1)

where:

I_t is the tabulated current-carrying capacity from Appendix 4 of BS 7671

C denotes a correction factor as follows:

C_a is ambient temperature; see Appendix C Table 4B1 (cables in air) or 4B2/4B3 (cables in ground)

C_g is grouping; see Appendix C Table 4C1 to 4C5

C_i is for conductors totally surrounded by thermal insulation. Although not given in Appendix 4 this factor is 0.5 (523.6.6).

It should be noted that some 17th Edition tables include a certain amount of thermal insulation and where these are used no further correction should be made. For example, new Table 4D5, included due to ECA suggestion, includes thermal insulation for cables in loft-roof spaces, and the C_i correction should not be used as it is already in the I_t value.

C_c is for semi-enclosed fuses to BS 3036.

A cable with tabulated current-carrying capacity (I_z) is then selected such to exceed the I_t (step 3). The procedure can be depicted as in Figure C 4.2.

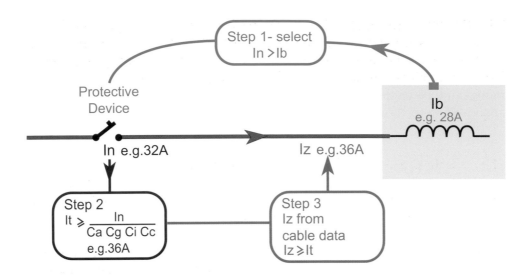

Figure C 4.2 Diagram of conductor size design method.

Current rating tables for some common cables are found in Appendix 2 of this guide. Correction factors are discussed in sections C 4.3.4 and C 4.3.5.

On completion of obtaining an installed current rating for the relevant cable size, it is then further checked for voltage drop, thermal withstand under fault conditions and size for earth fault loop impedance. These are discussed later in the chapter.

As the factors for thermal insulation and fusing (semi-enclosed) are not used all that often, Equation 1 becomes:

$$I_t = \frac{I_n}{C_a \, C_g}$$

(Equation 2)

The various correction factors are now discussed.

C 4.3.4 Grouping factors

Firstly, caution has to be exercised when applying grouping factors; very large cable sizes can result unless careful consideration of realistic design currents is made before a grouping factor is applied.

BS 7671: 2008 Appendix 4 includes revised information for grouping factors calculations. A new table of grouping factors for buried cables has been added (Table 4C2 reproduced below). Also new is a method for calculating the grouping factor for circuits where cables are of different sizes, as follows:

$$F = \frac{1}{\sqrt{n}}$$

where F is the group rating factor and n is the number of circuits.

This formula will give a lower group factor than BS 7671 Tables 4C1 to 4C5 as these tables assume that cables in the group are of the same size.

An overriding point to note before the grouping factor tables are given is the note 523.5 in BS 7671, which states that where a cable is known to carry 30% or less of its grouped rating, it can be ignored for the purposes of grouping. This is invaluable and should be utilized for BS 7671: 2008 calculations.

The procedure for applying grouping factors is either to use the factor from BS 7671 Tables 4C1 to 4C2 or to use the following method:

Compare the formula:

$$I_t = \sqrt{I_n^2 + 0.48I_b^2 \left(\frac{1 - C_g^2}{C_g^2}\right)} \qquad \text{(Equation 3)}$$

with

$$I_t = \frac{I_b}{C_g} \qquad \text{(Equation 4)}$$

using the larger value. This method can only be used where the circuits within a group are not expected to be simultaneously overloaded.

For circuits that are not fully loaded, the use of this (Equation 3) method produces a lower I_t than the grouping factor method using tables 4C1 to 4C5 from BS 7671. The procedures are summarized as:

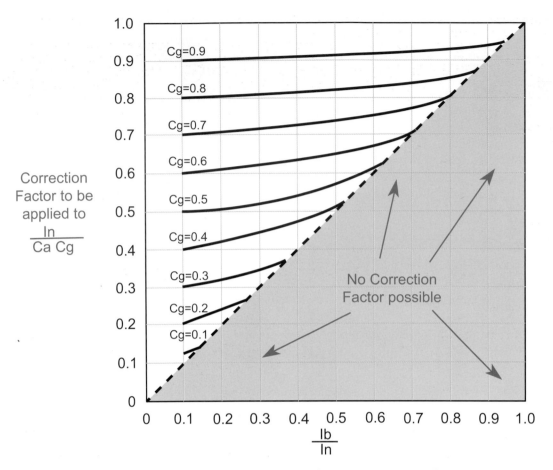

Figure C 4.3 Reduction formula for circuits not fully loaded. Source: adapted from *Electrical Installation Calculations*, *Third Edition*, ECA 2002.

- if a quick sizing method is required, use BS 7671 Tables 4C1 to 4C5 and equations 2 or 3 and do no more; or
- calculate I_t using Equations 3 and 4, using the larger I_t value.

As both C_g and I_b/I_n are both known, to make life even easier and to save application of the second bullet point above, which can be tiresome, the 'look-up' table in Figure C 4.3 can be used instead and applied to Equations 1 or 2 (but *not* where the protective device is a semi-enclosed fuse).

The following tables of grouping factors are reproduced from BS 7671: 2008.

Table 4C1 Rating factors for one circuit or one multicore cable or for a group of circuits, or a group of multicore cables, to be used with current-carrying capacities of Tables 4D1A to 4J4A

Arrangement (cables touching)	Number of circuits or multicore cables												To be used with current-carrying capacities, reference
	1	2	3	4	5	6	7	8	9	12	16	20	
Bunched in air, on a surface, embedded or enclosed	1.00	0.80	0.70	0.65	0.60	0.57	0.54	0.52	0.50	0.45	0.41	0.38	Methods A to F
Single layer on wall or floor	1.00	0.85	0.79	0.75	0.73	0.72	0.72	0.71	0.70	0.70	0.70	0.70	Method C
Single layer multicore on a perforated horizontal or vertical cable tray systems	1.00	0.88	0.82	0.77	0.75	0.73	0.73	0.72	0.72	0.72	0.72	0.72	Methods E and F
Single layer multicore on cable ladder systems or cleats, etc.	1.00	0.87	0.82	0.80	0.80	0.79	0.79	0.78	0.78	0.78	0.78	0.78	

NOTE 1: These factors are applicable to uniform groups of cables, equally loaded.

NOTE 2: Where horizontal clearances between adjacent cables exceeds twice their overall diameter, no rating factor need be applied.

NOTE 3: The same factors are applied to:
- groups of two or three single-core cables;
- multicore cables.

NOTE 4: If a system consists of both two- and three-core cables, the total number of cables is taken as the number of circuits, and the corresponding factor is applied to the tables for two loaded conductors for the two-core cables, and to the tables for three loaded conductors for the three-core cables.

NOTE 5: If a group consists of n single-core cables it may either be considered as $n/2$ circuits of two loaded conductors or $n/3$ circuits of three loaded conductors.

NOTE 6: The values given have been averaged over the range of conductor sizes and types of installation included in Tables 4D1A to 4J4A; the overall accuracy of tabulated values is within 5%.

NOTE 7: For some installations and for other methods not provided for in the above table, it may be appropriate to use factors calculated for specific cases; see for example Tables 4C4 and 4C5.

NOTE 8: When cables having differing conductor operating temperature are grouped together, the current rating is to be based upon the lowest operating temperature of any cable in the group.

NOTE 9: If, due to known operating conditions, a cable is expected to carry not more than 30% of its *grouped* rating, it may be ignored for the purpose of obtaining the rating factor for the rest of the group.

For example, a group of N loaded cables would normally require a group reduction factor of Cg applied to the tabulated It. However, if M cables in the group carry loads which are not greater than 0.3 CgIt amperes the other cables can be sized by using the group rating factor corresponding to (N-M) cables.

Table 4C2 Rating factors for more than one circuit, cables laid directly in the ground – Reference Method D in Tables 4D1A to 4J4A single-core or multicore cables

Number of circuits	Cable to cable clearance (α)				
	Nil (cables touching)	One cable diameter	0.125 m	0.25 m	0.5 m
2	0.75	0.80	0.85	0.90	0.90
3	0.65	0.70	0.75	0.80	0.85
4	0.60	0.60	0.70	0.75	0.80
5	0.55	0.55	0.65	0.70	0.80
6	0.50	0.55	0.60	0.70	0.80

Multicore cables

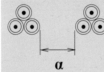

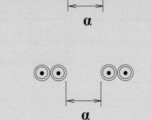

Single-core cables

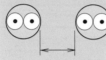

NOTE 1: Values given apply to an installation depth of 0.7 m and a soil thermal resistivity of 2.5 K.m/W. These are average values for the range of cable sizes and types quoted for Tables 4D1A to 4J4A. The process of averaging, together with rounding off, can result in some cases in errors of up to ±10%. (Where more precise values are required they may be calculated by methods given in BS 7769 (BS IEC 60287).)

NOTE 2: In case of a thermal resistivity lower than 2.5 K.m/W the correction factors can, in general, be increased and can be calculated by the methods given in BS 7769 (BS IEC 60287).

Table 4C3 Rating factors for more than one circuit, cables laid in ducts in the ground –
Reference Method D in Tables 4D1A to 4J4A

(i) Multicore cables in single-way ducts

Number of cables	Duct to duct clearance (α)			
	Nil (ducts touching)	0.25 m	0.5 m	1.0 m
2	0.85	0.90	0.95	0.95
3	0.75	0.05	0.90	0.95
4	0.70	0.80	0.85	0.90
5	0.65	0.80	0.85	0.90
6	0.60	0.80	0.80	0.90

Multicore cables

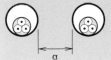

NOTE 1: Values given apply to an installation depth of 0.7 m and a soil thermal resistivity of 2.5 K.m/W.
They are average values for the range of cable sizes and types quoted for Tables 4D1A to
4J4A. The process of averaging, together with rounding off, can result in some cases in errors
of up to ±10%. (Where more precise values are required they may be calculated by methods
given in BS 7769 (BS IEC 60287).)

NOTE 2: In case of a thermal resistivity lower than 2.5 K.m/W the correction factors can, in general, be
increased and can be calculated by the methods given in BS 7769 (BS IEC 60287).

(ii) Single-core cables in non-ferrous single-way ducts

Number of cables	Duct to duct clearance (α)			
	Nil (ducts touching)	0.25 m	0.5 m	1.0 m
2	0.80	0.90	0.90	0.95
3	0.70	0.80	0.85	0.90
4	0.65	0.75	0.80	0.90
5	0.60	0.70	0.80	0.90
6	0.60	0.70	0.80	0.90

Single-core cables

NOTE 1: Values given apply to an installation depth of 0.7 m and a soil thermal resistivity of 2.5 K.m/W.
They are average values for the range of cable sizes and types quoted for Tables 4D1A to
4J4A. The process of averaging, together with rounding off, can result in some cases in errors
of up to ±10%. (Where more precise values are required they may be calculated by methods
given in BS 7769 (BS IEC 60287).)

NOTE 2: In case of a thermal resistivity lower than 2.5 K.m/W the correction factors can, in general, be
increased and can be calculated by the methods given in BS 7769 (BS IEC 60287).

Table 4C4 Rating factors for groups of more than one multicore cable, to be applied to reference current-carrying capacities for multicore cables in free air – Reference Method E in Tables 4D1A to 4J4A

Installation method in Table 4A2			Number of trays or ladders	Number of cables per tray or ladder					
				1	2	3	4	6	9
Perforated cable tray systems (Note 3)	31	Touching ⩾20 mm ⩾300 mm	1	See item 4 of Table 4C1					
			2	1.00	0.87	0.80	0.77	0.73	0.68
			3	1.00	0.86	0.79	0.76	0.71	0.66
			6	1.00	0.84	0.77	0.73	0.68	0.64
		Spaced De ⩾ 20 mm	1	1.00	1.00	0.98	0.95	0.91	—
			2	1.00	0.99	0.96	0.92	0.87	—
			3	1.00	0.98	0.95	0.91	0.85	—
Vertical perforated cable tray systems (Note 4)	31	Touching ⩾ 225 mm	1	See item 4 of Table 4C1					
			2	1.00	0.88	0.81	0.76	0.71	0.70
		Spaced ⩾ 225 mm De	1	1.00	0.91	0.89	0.88	0.87	—
			2	1.00	0.91	0.88	0.87	0.85	—
Unperforated cable tray systems	30	Touching ⩾20 mm ⩾300 mm	1	0.97	0.84	0.78	0.75	0.71	0.68
			2	0.97	0.83	0.76	0.72	0.68	0.63
			3	0.97	0.82	0.75	0.71	0.66	0.61
			6	0.97	0.81	0.73	0.69	0.63	0.58

(Continued.)

Table 4C4 *(Continued.)*

Installation method in Table 4A2			Number of trays or ladders	Number of cables per tray or ladder					
				1	2	3	4	6	9
Cable ladder systems, cleats, wire mesh tray, etc. (Note 3)	32 33 34	Touching ≥ 20 mm ≥ 300 mm	1 2 3 6	See item 4 of Table 4C1					
				1.00	0.86	0.80	0.78	0.76	0.73
				1.00	0.85	0.79	0.76	0.73	0.70
				1.00	0.84	0.77	0.73	0.68	0.64
		Spaced D_e ≥ 20 mm	1 2 3	1.00	1.00	1.00	1.00	1.00	—
				1.00	0.99	0.98	0.97	0.96	—
				1.00	0.98	0.97	0.96	0.93	—

NOTE 1: Values given are averages for the cable types and range of conductor sizes considered in Tables 4D1A to 4J4A. The spread of values is generally less than 5%.

NOTE 2: Factors apply to single layer groups of cables as shown above and do not apply when cables are installed in more than one layer touching each other. Values for such installations may be significantly lower and must be determined by an appropriate method.

NOTE 3: Values are given for vertical spacing between cable trays of 300 mm and at least 20 mm between cable trays and wall. For closer spacing the factors should be reduced.

NOTE 4: Values are given for horizontal spacing between cable trays of 225 mm with cable trays mounted back to back. For closer spacing the factors should be reduced.

Table 4C5 Rating factors for groups of one or more circuits of single-core cables to be applied to reference current-carrying capacity for one circuit of single-core cables in free air – Reference Method F in Tables 4D1A to 4J4A

Installation method in Table 4A2			Number of trays or ladders	Number of three-phase circuits per tray or ladder			Use as a multiplier to rating for
				1	2	3	
Perforated cable tray systems (Note 3)	31	Touching ≥ 300 mm ≥ 20 mm	1 2 3	0.98 0.96 0.95	0.91 0.87 0.85	0.87 0.81 0.78	Three cables in horizontal formation
Vertical perforated cable tray systems (Note 4)	31	Touching 225 mm	1 2	0.96 0.95	0.86 0.84	— —	Three cables in vertical formation

(Continued.)

Table 4C5 (*Continued.*)

Installation method in Table 4A2			Number of trays or ladders	Number of three-phase circuits per tray or ladder			Use as a multiplier to rating for
				1	2	3	
Cable ladder systems, cleats, wire mesh tray, etc. (Note 3)	32 33 34	Touching	1 2 3	1.00 0.98 0.97	0.97 0.93 0.90	0.96 0.89 0.86	Three cables in horizontal formation
Perforated cable tray systems (Note 3)	31		1 2 3	1.00 0.97 0.96	0.98 0.93 0.92	0.96 0.89 0.86	Three cables in trefoil formation
Vertical perforated cable tray systems (Note 4)	31	Spaced	1 2	1.00 1.00	0.91 0.90	0.89 0.86	
Cable ladder systems, cleats, wire mesh tray, etc. (Note 3)	32 33 34		1 2 3	1.00 0.97 0.96	1.00 0.95 0.94	1.00 0.93 0.90	

NOTE 1: Values given are averages for the cable types and range of conductor sizes considered in Tables 4D1A to 4J4A. The spread of values is generally less than 5%.

NOTE 2: Factors apply to single layer groups of cables (or trefoil groups) as shown above and do not apply when cables are installed in more than one layer touching each other. Values for such installations may be significantly lower and must be determined by an appropriate method.

NOTE 3: Values are given for vertical spacing between cable trays of 300 mm and at least 20 mm between cable trays and wall. For closer spacing the factors should be reduced.

NOTE 4: Values are given for horizontal spacing between cable trays of 225 mm with cable trays mounted back to back. For closer spacing the factors should be reduced.

NOTE 5: For circuits having more than one cable in parallel per phase, each three-phase set of conductors is to be considered as a circuit for the purpose of this table.

4.3.5 Other correction factors

As well as grouping factors, correction factors for ambient temperature (C_a), total surround by thermal insulation factor (C_i) and semi-enclosed fuse factor (C_c), which is 0.725) need to be used where appropriate.

Ambient temperature C_a

This is a factor for ambient temperature of the installation where the cables are run. The tabulated cable current ratings in Appendix 4 of BS 7671 are based upon a 30°C ambient temperature for general cables and 20°C around ambient temperature for cables buried in the ground. Tables 4B1 and 4B2 of Appendix 4 of BS 7671 give the relevant factors to be used, including the new correction factors for cables buried in the ground. Where there are mixed installation temperatures in a cable length it is best to use the higher temperature (alternatively, larger cables can be installed for that portion).

Thermal insulation C_i

A careful application of this factor is required, as many of the cable rating tables (4D to 4F) already allow for some thermal insulation. This factor ($C_i = 0.5$) is to be used where an appropriate cable installation method is not available (533.6.6). It must be applied to the 'in free air' rating of the cable type.

Semi-enclosed fuse factor C_c

This factor, equal to 0.725, should be used where the protective device is a semi-enclosed (rewirable) fuse.

C 4.3.6 Omission of overload protection

There are examples of circuits or equipment where it is recognized by BS 7671 that overload protection is not required or perhaps not desired.

The first of these is not utilized as much as it could be, and is for circuits with loads not likely to cause an overload (433.3.1). An example would be a circuit supplying a fixed water heater, but this regulation applies to many fixed loads. Fault current protection is still a requirement and must be provided.

The other category of loads for which overload protection need not be provided is where disconnection could cause danger, such as sprinkler pump supplies, lifting electromagnets, safety supplies in general (may not apply to all of them) and others.

It is noted that supply cut-out fuses are allowed to be utilized for overload protection of the main supply cables (433.3.1).

C 4.4   Fault protection

C 4.4.1 General and omissions

BS 7671 has new terminology for fault protection. Fault currents now include all faults, i.e. those between live conductors as well as earth faults. The term 'short circuit current' is not used within Section 434 although it still appears in part 2 – definitions (it was used in the 16th Edition).

There are two aspects of protection against fault currents: the protection of the cables, covered in this part, and the protection of and selection of equipment (fault rating), which is dealt with in Chapter D.

C 4.4.2 Fault protection requirements

For protection against short circuit, the overcurrent device must be able to:

● withstand the short circuit current (device breaking capacity); and
● disconnect sufficiently quickly to prevent damage to the cables.

The fault currents to be considered include faults between line conductors and earth, line conductor and neutral, and line-to-line conductors. The highest fault currents will arise with three-phase line conductors shorting together and to earth.

The BS 7671 requirements for fault protection are summarized in Table C 4.2.

Table C 4.2 Fault protection requirements.

Requirement	Regulation number
Fault current shall be determined at every position	434.1
Devices capable of withstanding fault levels	434.5.1
Disconnection times are to protect heat rise in cables[1]	434.5.2
Separate overload and fault devices to be co-ordinated	435.2

Note 1: Perhaps one of the most significant regulations here is in relation to achieving compliance with 434.5.2, which states that if overload protection ($I_b \leq I_n \leq I_z$) is achieved then this will provide fault protection (435.1).

C 4.4.3 Determination of fault current

The Regulations (434.1) require that the fault current be determined at every relevant point. For small and medium installations with LV utility supplies, this is satisfied by determination of the fault currents at the incomer. The DNO will supply this information and usually quote a maximum fault current of 16 kA. This is a theoretical maximum that is rarely found in practice. This level of fault will mean that MCBs (60898 devices, see Chapter D) will need to be rated at 16 kA. Whilst it is possible to purchase 16 kA devices, they are usually double width and often more expensive than the 3, 6, 9 or 10 kA devices. A solution here is to measure the incoming fault currents or loop impedances, and use these values.

For installations with private transformers, calculations are required, and this is outside the scope of this book. Generally, fault levels of 15–20 kA exist in the close vicinity of main switchgear, and a careful consideration of MCB fault rating must be made for local distribution boards.

C 4.4.4 Fault capacity of devices

Table C 4.3 summarizes fault capacities of relevant devices.

For most domestic installations the prospective fault current is unlikely to exceed 6 kA, up to which value the I_{cn} and I_{cs} values are the same.

The short-circuit capacity of devices to BS EN 60947–2 is specified by the individual manufacturer.

C 4.4.5 Circuits without overload protection

Where the overcurrent device does not provide overload protection, the cable size must be checked for short-circuit (thermal) withstand. This confirms that the circuit energy let through by the protective device does not cause a damaging heat rise in the cable. The calculation uses the 'adiabatic' equation given in Regulation 434.5.2 as follows:

$$t = \frac{k^2 S^2}{I^2}$$
(Equation 5)

where:

t is the duration in seconds

S is the cross-sectional area of conductor in mm^2

Table C 4.3 Fault ratings of common devices.

Device type	Fault current capacity (kA)
Circuit-breakers to BS EN 60898[1] and BS EN 61009[1]	I_{cn} I_{cs} 1.5 (1.5) 3.0 (3.0) 4.5 (4.5) 6 (6.0) (7.5) 16
Circuit-breakers to BS EN 60947–2	Varies, specified by manufacturer
Cartridge fuse to BS 1361 type I Type II	16.5 33.0
General purpose fuse to BS 88 Part 2.1 Part 6	50 at 400V 16.5 at 230V 80.0 at 400V
Semi-enclosed fuse to BS 3036 with category of duty	S1A-1 S2A-2 S4A-4
Circuit-breakers to BS 3871	M1–1 M1.5–1.5 M3–3 M4.5–4.5 M6–6 M9–9

Note 1: Two rated short-circuit ratings are defined in BS EN 60898 and BS EN 61009: I_{cn} is the rated short-circuit capacity (marked on the device), I_{cs} is the service short-circuit capacity. The difference between the two is the condition of the circuit breaker after manufacturer's testing. I_{cn} is the maximum fault current the breaker can interrupt safely, although the breaker may no longer be usable. I_{cs} is the maximum fault current the breaker can interrupt safely without loss of performance. The I_{cn} value is normally marked on the device in a rectangle, e.g.

```
6000
```

For the majority of applications the prospective fault current at the terminals of the circuit-breaker should not exceed this value.

I is the effective short-circuit current in amperes, due account being taken of the current limiting effect of the circuit impedances

k is a factor taking account of the resistivity, temperature coefficient and heat capacity of the conductor material, and the appropriate initial and final temperatures.

It should be noted that in this application of Equation 5, the disconnection time should not be taken as 5 s, but is taken from the time-current characteristic for the

Table C 4.4 *k* values for use in adiabatic equation.

| | 70° PVC | 90° XLPE | MICC | |
	(thermoplastic)	(thermosetting)	Served	Bare
Copper conductor	$k = 115$	$k - 100$	$k = 115$	$k = 115$
Aluminium conductor	$k = 76$	$k - 66$	N/A	N/A

protective device. Alternatively, some manufacturers provide I^2t values that can be used.

For copper and aluminium cables up to $300\,mm^2$, k is given in Table C 4.4; the full list for other materials is found in the Regulations.

The adiabatic equation (Equation 5) can be used to find a tripping time that will not overheat the cable being used. It is then compared with the actual tripping time for the protective device at the fault current level found in the device time/current curves. This can be a little confusing and it can be easier to calculate the minimum size required using a transposed adiabatic equation as follows:

$$S = \frac{I\sqrt{t}}{k},$$ noting that S comes out in mm^2.

It should be noted that where the initial cable size has been adjusted following a thermal withstand check, further iterations may be necessary as the new size itself affects the prospective fault current.

It is worth noting that the factors in Table 43.1 of BS 7671 are based on the circuit conductor running at its maximum operating temperature. Designers should take account of this when the adiabatic calculation yields a cable size 'just over' a standard available size. It is suggested that where this is in the order of 5% to 10%, the smaller size be used.

C 4.5   Voltage drop

C 4.5.1 7671 requirements
Section 525 of BS 7671: 2008 contains four regulations specifying that the voltage at the terminals of equipment is:

● suitable for that specified in the equipment product standard; or
● for equipment without a standard, suitable for safe functioning.

Table C 4.5 Appendix12 voltage drop specification.

Type of supply	Voltage drop lighting	Voltage drop other
From DNO	3%	5%
Private supply	6%	8%

A new Appendix (12) is included in BS7671: 2008 for consumers' installations which, if followed, is 'deemed to comply' with voltage drop design requirements of the Regulations. This is summarized in Table C 4.5.

It is recognized in the body of Section 525 and Appendix 12 that high inrush currents may cause higher voltage drop levels, and reference to equipment product standards is made. In essence this does not change the basic requirement that equipment must work!

C 4.5.2 System design values

For domestic installations, voltage drop is quite simple and rarely needs a design or further consideration.

In commercial and industrial installations, voltage drop design will depend upon the nature of the supply. If the supply is to the Electricity Safety, Quality and Continuity Regulations 2002, the supply voltage can vary between statutory limits of +10% to –6%. As such, the voltage drops in Table C 4.5 are recommended with the caveat that voltage can be lower if the equipment standard allows.

For private transformers the subject is more complicated, as the voltage regulation of the transformer is now under the control of you, the designer. The secondary terminal voltage of the transformer will largely depend upon the magnitude of the total load at any instant. Generally it is good practice to simulate the values of statutory voltage limits found in the Electricity, Safety, Quality and Continuity Regulations. The current limits are nominal voltage +10% to –6%. Thus for a large installation, a system of main and sub voltage drops may be required to be set, to achieve a 'system' design. A simple system view of this is given in Figure C 4.4.

A guide to installation voltage drop limits is 6% with an 8% maximum. Often it can be useful to apply values in stages in a system, and popular values are 2% to 3% for submains coupled with 4% or 3% for final circuits.

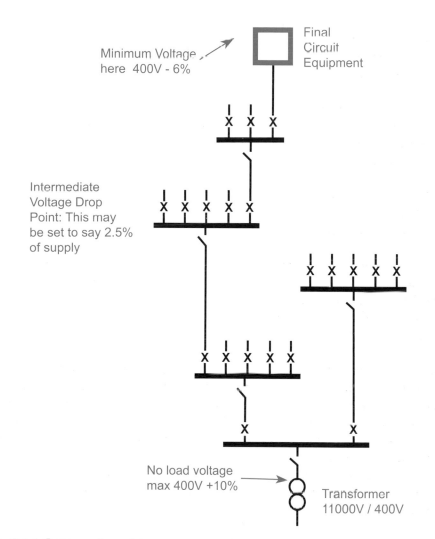

Figure C 4.4 System voltage drops.

C 4.5.3 Definition and simple calculation of voltage drop

An accurate estimation of the circuit design current is decisive in calculating an accurate voltage drop. For submains, diversity will need to be considered with a slightly different approach than when applying diversity for cable sizing. Whilst short time overload currents will not cause problems to a system, voltage dips below a certain value may cause equipment to stop or malfunction. Therefore a more rigorous approach to individual submain diversity is required. An 'overall' diversity factor, however, may be equal to that used for the maximum demand estimate.

The definition of voltage drop is the voltage difference between any two points of a circuit or conductor, due to the flow of current.

Voltage drop information for installation cables is given in BS 7671 Appendix 4 tables expressed in millivolts for a current of one amp for one metre of the cable. Hence:

$$\text{Voltage drop (V)} = \frac{\text{tabulated voltage drop} \times \text{design current (A)} \times \text{length}}{1000}$$

The tabulated ratings are also denoted as '(mV/A/m)' ratings and the above equation can be expressed as follows:

$$\text{v.d. (V)} = \frac{(\text{mV/A/m}) \times L \times I_b}{1000}$$

or can be rearranged to find a limiting circuit length:

$$\text{Length (m)} = \frac{\text{permitted v.d. (V)} \times 1000}{(\text{mV/A/m}) \times I_b}$$

C 4.5.4 Correction for conductor temperature

These equations and the (mV/A/m) values in Appendix 4 of BS 7671 are based upon fully loaded circuits, a rare circumstance in practice.

The values of (mV/A/m) given in the Appendix 4 tables are at the maximum conductor operating temperature of, say, 70°C or 90°C, and these temperatures are only reached when the conductor is carrying its full load. At lower loads, the temperature and the resistance of the cable are lower. The difference can be significant – up to 20%. Thus for lightly loaded circuits this can reduce the tabulated rating (mV/A/m) by up to 20%.

The conductor temperature correction factor, C_t, is worked out using the following:

$$C_t = \frac{230 + t_p - \left(C_a^2 C_g^2 - \frac{I_b^2}{I_t^2} \right)(t_p - 30)}{230 + t_p}$$

where

t_p is the rated maximum conductor operating temperature

I_b is design current of the circuit

I_t is the tabulated current rating of the cable.

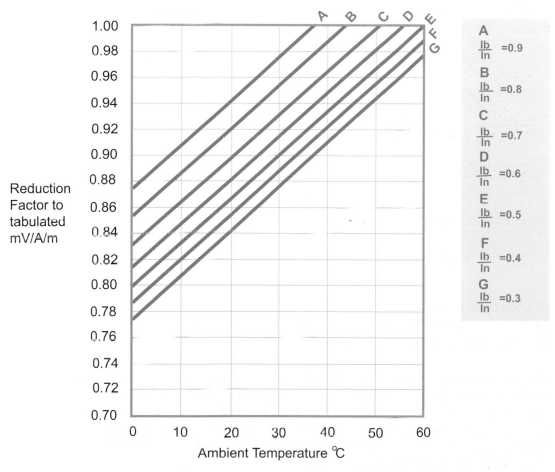

Figure C 4.5 Reduction factors for thermoplastic cables (PVC). Source: adapted from *Electrical Installation Calculations, Third Edition*, ECA 2002.

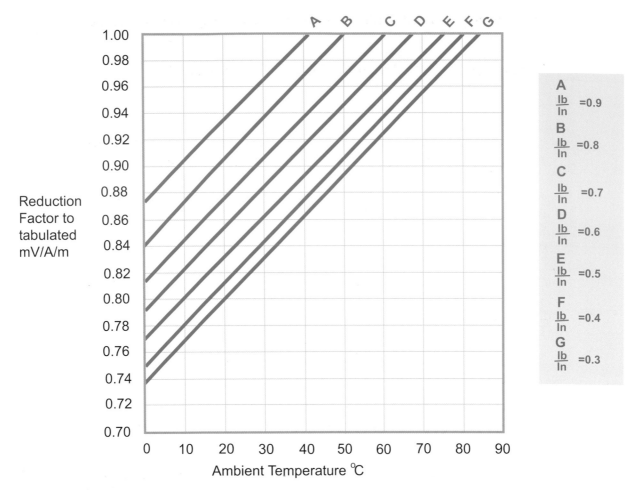

Figure C 4.6 Reduction factors for thermosetting cables (XLPE). Source: adapted from *Electrical Installation Calculations*, *Third Edition*, ECA 2002.

Again, manual calculations can be tedious, and the graphs in Figures C 4.5 and C 4.6 provide a quick and convenient way of avoiding them. The graphs can be used to correct tabulated (mV/A/m) values and can be applied to all cables under 16 mm² and to the tabulated resistive component, $(mV/A/m)_r$, for larger cables.

To use the graphs, select an I_b/I_n line, and find the reduction factor F at the appropriate installation ambient temperature and a corresponding reduction factor. This factor is multiplied by either the tabulated (mV/A/m) value for cables up to 16 mm² or by the resistive $(mV/A/m)_r$ component for cables of 25 mm² and over.

C 4.5.5 Correction for load power factor and temperature
For larger cables above 25 mm² where the load power factor is known, a more accurate estimation can be made of voltage drop allowing slightly greater cable

length or perhaps reduced size in marginal cases. The voltage drop is calculated using the formula:

$$\text{Voltage drop} = \frac{L \times I_b}{1000} [C_t \cos \phi \, (\text{mV/A/m})_r + \sin \phi \, (\text{mV/A/m})_x]$$

where

$(\text{mV/A/m})_r$ is the tabulated value of resistive element of voltage drop in mV per amp per metre from the cable rating tables of Appendix 4 of BS 7671

$(\text{mV/A/m})_x$ is the tabulated value of inductive element of voltage drop in mV per amp per metre from the cable rating tables of Appendix 4

ϕ is the power factor of the load

For cables of $25 \, \text{mm}^2$ and greater, the voltage drop is given both as an impedance $(\text{mV/A/m})_z$ and as a complex impedance with a resistive element $(\text{mV/A/m})_r$ and inductive element $(\text{mV/A/m})_x$ as follows:

$$(\text{mV/A/m})_z = \sqrt{(\text{mV/A/m})_r^2 + (\text{mV/A/m})_x^2}$$

Example calculations are provided in Appendix 10.

C 4.6   Disconnection and electric shock

C 4.6.1 Introduction and protective measures

A significant change for BS 7671: 2008 is the introduction of new terminology within the new Chapter 41. The previously very familiar terms 'direct contact' and 'indirect contact' are replaced by the terms 'basic protection' and 'fault protection' respectively; these terms in themselves introduce no technical changes.

As well as terminology changes, the whole of Chapter 41 has been revised. It is important to become familiar with the structure of the chapter, and this is depicted in Figure C 4.7.

It can be seen that 'basic protection' measures have been shunted to the end of the chapter. This is sensible, as almost without exception we do not consider these; they are so fundamental as to be 'automatically' included. The 'basic protection' protective measure is achieved by selecting equipment complying with relevant product standards.

The new terms 'basic protection' and 'fault protection' are illustrated in Figure C 4.8.

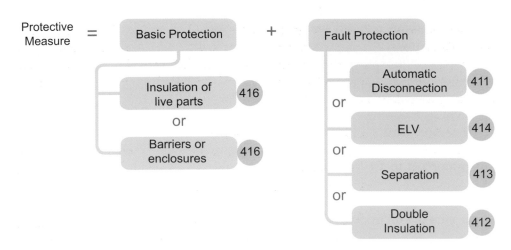

Figure C 4.7 Structure and 'protective measures' of Chapter 41 for general application.

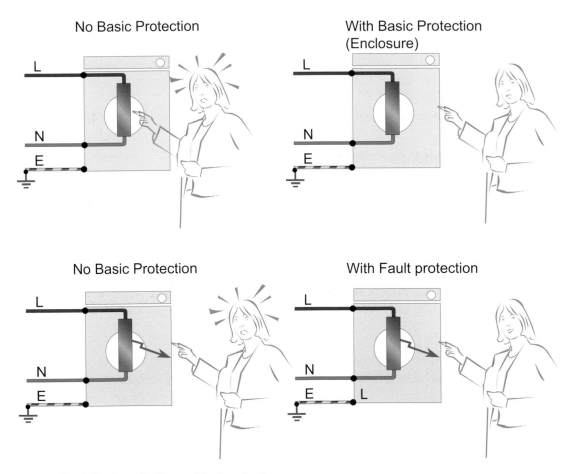

Figure C 4.8 Basic protection and fault protection.

The second structural arrangement of Chapter 41 to note is that the measures other than automatic disconnection or 'other measures', i.e. Obstacles and Out of Reach (417), Non-conducting Locations (418) and Earth-free Local Equipotential Bonding (418.2) have been relegated to the rear of the chapter. As these methods are rarely used they have been discussed in Appendix 13.

C 4.6.2 Automatic disconnection of supply

The 'protective measure – automatic disconnection of supply' is the standard method used in most applications and almost by default. This measure makes up the bulk of Chapter 41 totalling some 40 regulations.

Although the whole of Chapter 41 has been revised, highlights of automatic disconnection include reduced disconnection times (compared with the 16th Edition) and an increased use of RCDs, both explained in this part of the book.

Chapter 41 first sets down the main requirements for automatic disconnection, followed by specifics for TN, TT and IT systems.

To comply with the 'protective measure', the following are required:

1 basic protection; and
2 protective earthing and protective equipotential bonding (see Chapter E); and
3 automatic disconnection in the event of a fault.

Automatic disconnection requires disconnection within the time given by Table 41.1 of BS 7671. For nominal line-to-earth voltages U_0 of 230 V, as in the UK, the requirements are given in Table C 4.6.

Table C 4.6 Maximum disconnection times for U_0 230 V.

System	Final circuits disconnection time (s)	Distribution circuits disconnection time (s)
TN a.c.	0.4	5
TN d.c.	5	5
TT a.c.	0.2	1
TT d.c.	0.4	1

These disconnection times are achieved by limiting the earth fault loop impedance Z_s such that

$$Z_s \leq \frac{U_0}{I_a}$$

where

U_0 is the voltage to earth (normally 230 V)

I_a is the current in amperes causing the automatic operation of the disconnecting device within the time specified in the table

Z_s is the total earth fault loop impedance (ELI) in ohms of the fault loop, comprising the source impedance Z_e, the line conductor up to the point of the fault Z_1 and the protective conductor within the installation Z_2 between the point of the fault and the source, or

$$Z_s = Z_e + (Z_1 + Z_2)$$

Earth fault loop impedance calculations conventionally assume a fault of negligible impedance.

Figures C 4.9 and C 4.10 show the path of earth fault current for TN-C-S and TT systems and the components that make up the impedance of the earth fault loop.

For TT systems Z_s includes the resistance of the installation cables (Z_1 and Z_2), the installation earth electrode (Z_a), the supply distribution cable and the supply earth electrode (Z_d).

Maximum design earth fault loop impedance values for common devices are given in Tables 41.2, 41.3, 41.4 of BS 7671: 2008 and are reproduced in Appendix 3 of this book.

Within this book, in order to distinguish the BS 7671 tabulated design earth fault loop impedance values from measured values, design earth fault loop impedance values have been given the symbol Z_{41} (note: this symbol does not appear in BS 7671).

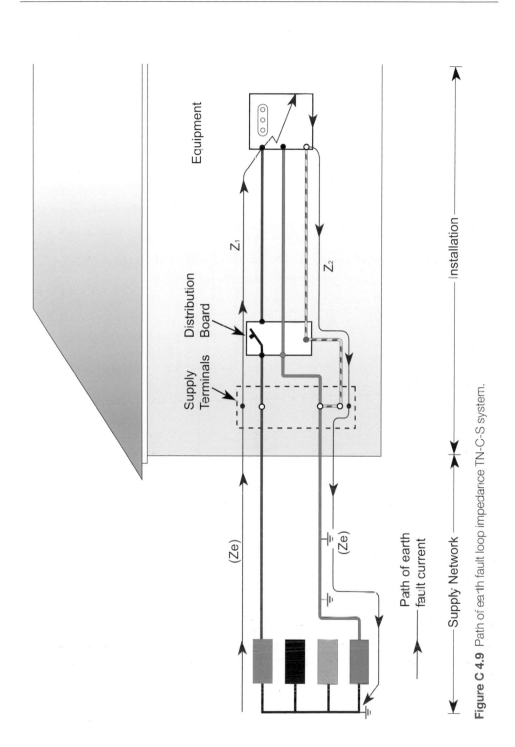

Figure C 4.9 Path of earth fault loop impedance TN-C-S system.

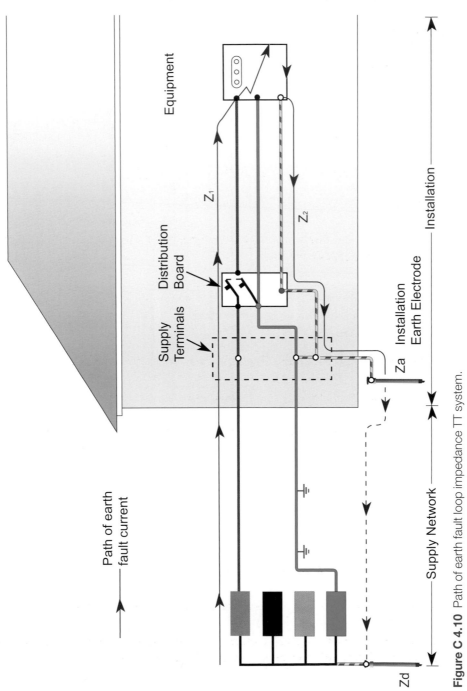

Figure C 4.10 Path of earth fault loop impedance TT system.

Therefore, to achieve the required disconnection times:

$$Z_{41} > Z_s \text{ and}$$

$$Z_{41} > Z_e + (Z_1 + Z_2)$$

where

Z_{41} is the design ELI from Chapter 41

Z_s is the total earth fault loop impedance

Z_e is the impedance of the source or sources.

For conductor sizes 16 mm^2 and less, reactance is not a factor and resistance can be used, and the equation becomes:

$$Z_{41} > Z_e + (R_1 + R_2)$$

where

R_1 is the resistance of the line conductor, up to the point of the fault

R_2 is the resistance of the protective conductor.

An alternative way of expressing this is by calculating the maximum length of a circuit and using the following formula:

$$L \leq \frac{Z_{41} - Z_e}{R_1 + R_2}$$

where

R_1 is the resistance in ohms per metre of the line conductor

R_2 is the resistance in ohms per metre of the protective conductor

Z_e, the impedance of the supply, is either measured, determined from the supply company, or if domestic can be assumed to be as in Table C 4.7.

Table C 4.7 Z_e for utility supplies.

Type of system	External earth fault loop impedance Z_e (Ω)
TN-C-S	0.35
TN-S	0.8
TT	21

C 4.6.3 TT installations

A disconnection time of 0.2 s for TT installations has been introduced in the 17th Edition. This makes the use of an RCD or circuit breaker with a residual element virtually essential.

For most installations, the installation earth electrode value Z_a (see Figure C 4.10) is large compared with the other factors contributing to earth fault loop impedance (supply and installation cables and supply earthing, Z_d). For simplicity, therefore, the following formula is used:

$$Z_a < Z_{41\ RCD}$$

where $Z_{41\ RCD}$ is the maximum value of loop impedance from Table C 4.8.

When an RCD is used for shock protection it is necessary to provide short circuit protection between line conductors and neutral by an overcurrent device.

Table C 4.8 TT maximum ELI values

RCD rated residual operating current (mA)	Maximum values of earth fault loop impedance $Z_{41\ RCD}$ (Ω)
30	1533[1]
100	460[1]
300	153
500	100

[1] An earth electrode value of greater than 200 Ω is considered by some to be unstable; others use a figure of 100 Ω. However, for single vertical rod electrodes, depth is an important factor and a 2.4 m deep electrode with initial resistance of, say, 300 Ω will be more stable than a shallow 1.2 m rod with an initial resistance of, say, 100 Ω. Thus the effects of freezing and drying out can be all but eliminated in the UK by installing 2.4 m deep electrodes.

C 4.6.4 ELI adjustment

The design values of maximum earth fault loop impedance (ELI) given in BS 7671 Tables 41.2, 41.3 and 41.4 cannot be directly compared to measured values. Measured values of ELI need to be adjusted to take account of the fact that when measured they are not at maximum operating temperature. While this point is discussed in Chapter F of this book, the necessary compensation step often gets missed; the design engineer thinks the tester will do a compensation and vice versa.

A new appendix, Appendix 14, has been included in the 17th Edition, and this suggests a compensation value of 0.8 applied as follows:

$$Z_{measured} \leq 0.8 \, Z_{41}$$

Although this guidance and the values in BS 7671 Tables 41.2, 41.3 and 41.4 set figures for maximum ELI, we should not get paranoid if the odd circuit slightly exceeds these values.

C 4.6.5 Irrelevant ELI specification

A common source of misunderstanding is that of either specifying or measuring values of ELI where the circuit also has an RCD fitted. ELI measurement under these circumstances is a futile exercise. The circuit will have been checked for continuity, and this is all that is needed together with, of course, functional checks of the RCD. This criterion satisfies requirements for automatic disconnection. The subject is somewhat confused by the inclusion of RCBOs in BS 7671 Table 41.3 and for clarification, circuits with RCBOs do not need to meet the specified ELI values.

Functional checks for RCDs are covered in Chapter F.

C 4.6.6 RCD supplementary protection

A completely new requirement in Chapter 41 is Regulation 411.3.3, which specifies RCDs for socket-outlet circuits intended for use by ordinary persons.

A point to stress here is that the 'additional protection' is not optional; it is now a fundamental requirement of Chapter 41.

The wording of this regulation is reproduced below, and guidance on its interpretation is given under the selection and erection part of RCDs, Section D 7.1.

411.3.3 Additional Protection

In a.c. systems, additional protection by means of an RCD in accordance with Regulation 415.1 shall be provided for:

(i) socket-outlets with a rated current not exceeding 20 A that are for use by ordinary persons and are intended for general use, and

(ii) mobile equipment with a current rating not exceeding 32 A for use outdoors.

An exception is permitted for:

(a) socket-outlets for use under the supervision of skilled or instructed persons, e.g. in some commercial or industrial locations, or

(b) a specific labelled or otherwise suitably identified socket-outlet provided for connection of a particular item of equipment.

Note 1: See also Regulations 314.1(iv) and 531.2.4 concerning the avoidance of unwanted tripping.

Note 2: The requirements of Regulation 411.3.3 do not apply to FELV systems according to Regulation 411.7 or reduced low voltage systems according to Regulation 411.8.

In the context of this Regulation, RCD is a 30 mA RCD.

C 5 Submains

C 5.1 Diversity

Getting a design that achieves function, economics and future capacity is important for most installations and in achieving this an accurate estimate of design current is key. General notes on diversity and load profiles were given in Section C 3.

C 5.2 Distribution circuit (submain) selection

Distribution circuits are normally selected from the following cables and wiring systems:

Table C 5.1 Comparison of distribution systems.

Distributor system	Disadvantages	Advantages
Unarmoured cables with or without cable tray	Risk of damage	Quick to install
Armoured cables	Large sizes require support to supporting structures	Good mechanical protection, good fire withstand
Mineral insulated cables	Need installation skills, may need protection from voltage surges	Robust, good fire withstand
Single insulated in conduit or trunking	Limited current carrying capacity	Good mechanical protection, easy rewiring
Busbar trunking	Not easily adapted	Easy to alter tap-offs

- busbar trunking;
- single insulated cables in trunking or conduit;
- steel wire armoured cables;
- mineral insulated cables.

The most suitable can be selected with assistance from Table C 5.1, which generally highlights the particular advantages and disadvantages of the systems. Of course, regional material and labour costs will have to be factored into this decision of wiring system.

C 5.3 Armouring as a cpc

Software design packages may propose that a supplementary protective conductor be run in parallel with an armoured cable. This is usually proposed to reduce the loop impedance such that disconnection will occur within the required five seconds. If possible, designers should manually check the calculation at this point.

The armouring of cables can and should be used as a protective conductor.

All protective conductors are required either to:

- comply with the adiabatic equation of Regulation 543.1.3,

$$S = \frac{\sqrt{I^2 t}}{k}, \text{ or}$$

- meet the cross-sectional area requirements of Table 54.7.

The calculated size virtually always produces a smaller conductor size (see Chapter E).

In 2007, the ECA commissioned the Electrical Research Association (ERA) to carry out an investigation into the merits or not of running additional cpcs externally to armoured cables.

The report included a comparison of the fault current withstand of the armour of cables to BS 5467, using the k values given in Chapter 54 of BS 7671. It showed that for all cables except 120 mm^2 and 400 mm^2 2-core cables, the fault current-withstand of the armour was greater than the fault current required to operate a BS 88 fuse within 5 s. The 120 mm^2 cable was within a few per cent of passing, and should be used unless it is intended to run the cable at full current-carrying capacity.

Also, calculation of the current sharing between the armour and an external cpc showed that if a small external cpc was run in parallel with the armour of a large cable there is a risk that the fault current withstand of the external cpc will be exceeded. Because of this it is recommended that the cross-sectional area of the external cpc should not be less than a quarter of that of the line conductors.

More details of the findings from this report are discussed in Chapter E, Section E5, and a full copy of the report is included in Appendix 16.

The following guidance is recommended:

- Use all armoured cables as a cpc without the need for a thermal withstand check with the exception of 400 mm^2 2-core.
  Note: The 120 mm^2 cable was within a few per cent of the size required (required I^2t – 33 062 500, actual 33 800 000 after 5 s) and should be used without calculation unless it is intended to run the cable at full current-carrying capacity. It is very difficult to load a cable of this size at its full current-carrying capacity for a continuously long period.
- Where calculation software specifies external cpcs for ELI reduction, use a cpc size of at least 25% of the phase size (perhaps after manually checking calculations).
- Manually check ELI calculations where they specify a small external cable cpc.

Table C 5.2 Distribution circuit disconnection times.

System	Disconnection time for distribution circuits (s)
TN	5
TT	1

C 5.4   Automatic disconnection for submains

The disconnection times allowed for distribution circuits and for any circuit, other than a final circuit not exceeding 32 A, are given in Table C 5.2.

C 6   Discrimination co-ordination

C 6.1   Principles and system co-ordination

Section 536 of BS 7671: 2008 details requirements for protective device co-ordination. An overall 'system' design view has to be taken on discrimination and co-ordination as otherwise this can lead to uneconomic schemes.

For some installations, depending upon the number of protective devices between final circuit and incomer protection, it may be an expensive luxury to design for full discrimination. Consider Figure C 6.1, which illustrates this point; a final circuit distribution board has a socket outlet protected by a 32 A circuit breaker.

As can be seen, by using a 2:1 discrimination rule to achieve full discrimination, a 2000 A main protective device is required and we have not considered any loads!

This demonstrates that, for many installations, whole system discrimination is not justified unless there are life-critical constraints.

Regulation 536.1 states that discrimination should be considered to prevent danger and where required for proper functioning of the installation. Chapter 36 gives an example of life-support systems where discrimination should be considered.

There is a fair bit of judgement to be made. Take Figure C 6.1 again — let's say the final circuit board was for a multi-million pound dealer unit. It would be reasonable for circuit breaker A to discriminate with circuit breaker B, but the discrimination between circuit breakers C and D may still be regarded as a luxury. We all know that despite the turnover of the client, there would be a limit to what they would pay for the electrical installation.

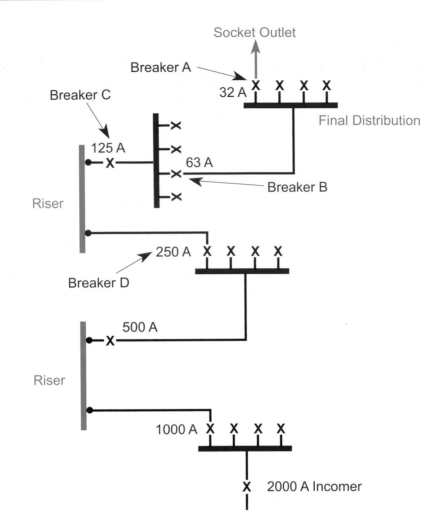

Figure C 6.1 System discrimination.

There are many options and combinations that can assist; for example, in Figure C 6.1 the final circuit breaker (A) could be 16 or 20 A, or circuit breaker C may be omitted.

In order to carry out an accurate discrimination study, fault current magnitudes are required at every protective device position.

C 6.2   Fuse-to-fuse discrimination

Fuses are relatively simple to consider when establishing discrimination co-ordination. As a rule of thumb, an upstream circuit breaker will always discriminate with a downstream fuse of half its rating. This applies for all currents of both overload magnitude and for fault currents. Fuses have characteristics described as follows:

- Pre-arcing I^2t is the energy required to take the fuse element to the point where it starts to melt.
- Total operating I^2t is the total energy until the arc is quenched.

This information is widely available from product standards, manufacturers and is included in Appendix 5 of this book. Discrimination is achieved if the pre-arcing I^2t of the upstream device is less than the total operating I^2t of the downstream device; this is illustrated for some common BS 88 fuses in Figure C 6.2.

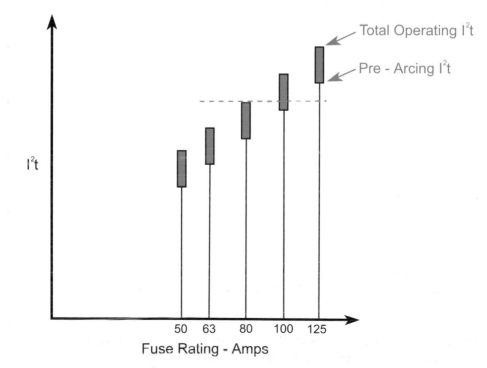

Figure C 6.2 Typical fuse I^2t characteristics.

It can be seen that by studying the I^2t characteristics that the 80 A fuse discriminates with the 125 A fuse (but not with the 100 A fuse).

Fuses to BS 88 and BS 1361 have relatively very high fault ratings and will be able to interrupt all short circuit currents in the installation, with no need for cascading (see Section C 6.3).

C 6.3   Circuit breaker to circuit breaker discrimination

Unlike fuses, circuit breakers cannot always rely on the upstream/downstream ratio of 2:1 for guaranteed discrimination. The 2:1 ratio generally holds true for the thermal trip mechanism but does not always apply to the magnetic trip mechanism (see D5 for explanation of mechanisms).

The normal method of manual co-ordination is by plotting the relevant protective device characteristics onto the same time-current graph paper. Due to scaling this is carried out on log-log paper; the graph in Figure C 6.3 shows two circuit breakers and the 'zone' where discrimination may not be achieved.

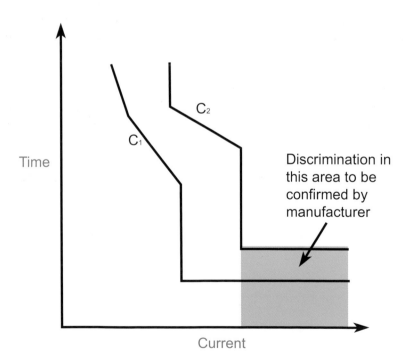

Figure C 6.3 Discrimination between two circuit breakers.

The zone indicated in Figure C 6.3 indicates an area where discrimination can only be achieved by carrying out tests and thus only manufacturers can confirm discrimination between circuit breakers in this zone.

Discrimination in this area is a subject vaguely addressed by some manufacturers and, for critical applications mentioned in C 6.1, confirmation by the manufacturer is strongly recommended.

Sometimes this subject is confused with cascading, a circumstance where downstream circuit breakers are 'backed-up' by circuit breakers. This technique should not be a substitute for a system discrimination co-ordination. Where cascading is used, the upstream breaker must have sufficient short-circuit breaking capacity to interrupt the fault, and the downstream breaker must be able to handle the through fault currents sufficiently long enough for the upstream breaker to operate.

C 6.4 Circuit breaker to fuse discrimination

At certain fault levels, fuses operate quicker than circuit breakers. Although now always readily available, the use of fuses is often to assist with achieving a system discrimination scheme. Figure C 6.4 shows circuit breaker and fuse characteristics and indicates the faster operating time at high fault currents (no values have been put on the figure as this is a general trend).

The principle outlined in Figure C 6.4 is the same as used for back-up protection, where a downstream circuit breaker is not rated for the prospective fault current at the point where it is installed. Take point X in the figure; larger fault currents will be interrupted by the fuse, smaller fault currents will be interrupted by the circuit breaker. Thus point X should be at a current less than the fault rating of the circuit breaker. Manufacturers' data should be considered in co-ordinating for back-up protection.

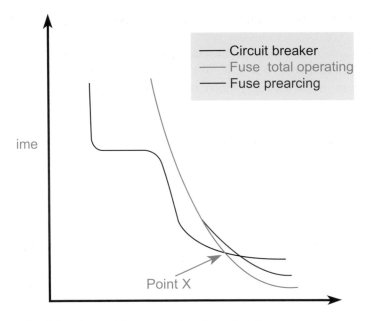

ime

Point X

Figure C 6.4 Circuit breaker and fuse operating characteristics.

C 7 Parallel cables

C 7.1 General and 7671 requirements

The use of parallel cables is often made for reasons of cost as, for a given current, two or more cables run in parallel may be more economic than a single larger one.

The other reason to use two or more cables in parallel is for ease of running and the logistics of termination. Sometimes switchgear may simply not have the space in the termination area for a single cable over a certain size.

BS 7671 requirements for cables run in parallel are summarized in Table C 7.1.

Table C 7.1 Requirements for cables in parallel.

Requirement	Regulation
No branch circuits allowed	433.4
$I_{z\,total} = I_{z1} + I_{z2} \ldots$ etc.	433.4.1
Unequal size conductors require careful consideration (see C 7.2)	433.4.2
Fault protection shall be at supply end	434.4
Shall have equal current sharing or calculated appropriately	523.8

It should be noted that ring circuits are not considered to be parallel cables.

The requirements of Table C 7.1 and hence the Regulations are satisfied by selecting cables of exactly the same type, construction and installation, in which case the total current-carrying capacity is the sum of the individual current-carrying capacities.

It should be noted that grouping factors are applicable to parallel cables.

C 7.2 Unequal current sharing

The general rules, summarized in Table C 7.1, apply to most applications for cables run in parallel with equal impedance. This is achieved if all the factors that affect the current-carrying capacity of the cables are the same, i.e. type, length, construction and installation method. It is assumed by this criterion that the impedances will be within 10% of each other and current sharing is taken to be equal.

Where this is not the case, it is possible to calculate and design the unequal current sharing based upon the different impedances, and methods for this are included in Appendix 10 of BS 7671: 2008. Discussion of this is outside the scope of this book.

 C 8 **Harmonics**

C 8.1 Requirements

BS 7671: 2008 primarily concerns itself with harmonics in respect of appropriate sizing of the neutral conductor, and the requirements are summarized in Table C 8.1.

Table C 8.1 Requirements for harmonic assessment for neutral conductor.

Requirement	Regulation	Notes
Balanced polyphase circuits require no action	523.6.1	
Neutral in unbalanced polyphase circuits to be sized for highest line current	523.6.2	This is usually done as a matter of course
For harmonic content > 15% reduced size neutral shall not be used	523.6.3	This is usually done as a matter of course
For high harmonic content further consideration is to be made	523.6.3	

Table C 8.2 Rating factors for triple harmonic currents in four-core and five-core cables.

Third harmonic content of line current[1] (%)	Rating factor	
	Size selection is based on line current	Size selection is based on neutral current
0–15	1.0	—
15–33	0.86	—
33–45	—	0.86
> 45	—	1.0

[1] Expressed as % of THD.

C 8.2   Harmonic assessment

Coupled with Regulation 523.6.3, BS 7671: 2008 provides an informative Appendix (11) on cable de-rating for installations with high harmonic currents.

Most modern installations contain equipment that causes harmonic currents; for example, equipment with switched mode power supplies found in personal computers, photocopiers, printers, fluorescent lighting and motor drives. The summation of these harmonic currents can cause an overload in the neutral conductor. Table C 8.2, extracted from Appendix 11 of BS 7671, provides de-rating factors for balanced three-phase loads and should be applied to 4- and 5-core cables.

It can be seen that a de-rating factor of 0.86 will cover neutral undersizing and perhaps may be used for loads like micro and small data centres and communications rooms. This de-rating should be used in conjunction with the general cable sizing philosophy given in Sections C 3 and C 4.

Appendix 11 of BS 7671: 2008 gives further methods for harmonic de-rating factors for higher harmonic frequencies, but these are outside the scope of this book.

C 9   Standard final circuit designs

C 9.1   Introduction and scope

This section provides 'standard' designs for some common final circuits. The designs provide circuit type, protective device and most importantly a maximum circuit length. Many of us use these standard designs, as it is time-consuming to design every final circuit from first principles. Standard circuit designs are usually used for common circuits, particularly domestic circuits.

The designs in this Section C 9.3 can be used in commercial and industrial installation designs for final circuits where the eath fault loop impedance at the local distribution board does not exceed the standard values given below.

The tables have been calculated using particular parameters as follows:

External earth fault loop impedances
Those used by the UK supply industry are as follows:

- $0.35\,\Omega$ for PME supplies;
- $0.80\,\Omega$ for TN-S supplies;
- $21\,\Omega$ for TT supplies.

Where RCDs are used, the tables in this section have been calculated so that the earth fault loop impedance values comply with the limiting values for overcurrent devices in Chapter 41. This gives maximum lengths, which are generally recommended, but considerably longer lengths are possible.

Cable types and installation methods
The standard circuit tables have assumed flat twin and earth cable to BS 6400 with reduced cpc size and installed as Table 4A2 of BS 7671 with current rating taken from Table 4D5 of BS 7671 (see Table C 9.1):

Table C 9.1 Flat twin and earth cable installation methods, adapted from Table 4D5 of BS 7671.

Reference method	Examples	Description	Rating of 2.5 mm² cable
A		Enclosed in conduit in an insulated wall	20

(Continued.)

Table C 9.1 (*Continued.*)

Reference method	Examples	Description	Rating of 2.5 mm² cable
C		Clipped direct	27
Method 100		Installation methods for flat twin and earth cable clipped direct to a wooden joist above a plasterboard ceiling with a minimum U value of 0.1 W/m²K and with thermal insulation not exceeding 100 mm in thickness	21
Method 101		Installation methods for flat twin and earth cable clipped direct to a wooden joist above a plasterboard ceiling with a minimum U value of 0.1 W/m²K and with thermal insulation exceeding 100 mm in thickness	17
Method 102		Installation methods for flat twin and earth cable in a stud wall with thermal insulation with a minimum U value of 0.1 W/m²K with the cable touching the inner wall surface	21
Method 103	with a current rating factor of 0.5 in accordance with Regulation 523.7	Installation methods for flat twin and earth cable in a stud wall with thermal insulation with a minimum U value of 0.1 W/m²K with the cable not touching the inner wall surface	13.5

Note that Table 4D5 was expanded for BS 7671: 2008, and confirms that in many cases cables installed in particular circumstances with thermal insulation (methods 100, 101 and 102) show a significantly better current rating as compared with the 'totally surrounded method' (method 103).

C 9.2   Standard domestic circuits

The following tables are based on a voltage drop of 5% generally and lighting circuit voltage drop of 3%.

Table C 9.2 Standard domestic final circuit lengths type B circuit breakers.

Type of circuit	Cable csa pvc/pvc (mm²)	Circuit breaker rating (A)	Maximum length for TN-C-S PME (m)		Maximum length for TN-S (m)		Installation method
			MCB	RCD or RCBO	MCB	RCD or RCBO	
Ring 13 A socket outlet	2.5/1.5	32	note 1	106	note 1	106	A, C, 100, 102; see note 1
Radial 13 A socket outlets	2.5/1.5	20	40	40	40	40	
Cooker (oven and hob)	6/2.5	32	53	53	50	53	
Oven (no hob) up to 3 kW	2.5/1.5	16	55	55	55	55	All
Immersion heater	2.5/1.5	16	55	55	55	55	
Shower to 30 A (7.2 kW)	6/2.5	32	note 1	53	note 1	50	A, C, 100, 102; see note 1
Shower to 40 A (9.6 kW)	10/4	40	note 1	66	note 1	66	
Storage heater	2.5/1.5	16	51	51	51	51	All
Fixed lighting (excl. switch drops)	1.5/1.0	10	53	53	53	53	

Note 1: If the cable is installed as per installation methods 101 or 103 a larger csa cable is required or thermosetting cable per 4E3A.
Note 2: RCD (or RCBO) required.

Table C 9.3 Standard domestic final circuit lengths type C circuit breakers.

Type of circuit	Cable csa pvc/pvc (mm²)	Circuit breaker rating (A)	Maximum length for TN-C-S PME (m)		Maximum length for TN-S (m)		Installation method
			MCB	RCD or RCBO	MCB	RCD or RCBO	
Ring 13 A socket outlet	2.5/1.5	32	63	82	NP	NP	A, C, 100, 102; see note 1
Radial 13 A socket outlets	2.5/1.5	20	34	40	14	19	
Cooker (oven and hob)	6/2.5	32	29	49	NP	NP	

(Continued.)

Table C 9.3 (Continued.)

Type C circuit breaker and RCBO (to BS EN 60898 or 61009)							
Type of circuit	Cable csa pvc/pvc (mm²)	Circuit breaker rating (A)	Maximum length for TN-C-S PME (m)		Maximum length for TN-S (m)		Installation method
			MCB	RCD or RCBO	MCB	RCD or RCBO	
Oven (no hob)	2.5/1.5	16	46	55	27	35	-
Immersion heater	2.5/1.5	16	46	55	27	35	A, C, 100, 102; see note 1
Shower to 30 A (7.2 kW)	6/2.5	32	note 2	49	NP	NP	
Shower to 40 A (9.6 kW)	10/4	40	note 2	51	NP	NP	All
Storage radiator	2.5/1.5	16	46	51	21	35	
Fixed lighting (excl. switch drops)	1.5/1.0	6	53	53	53	53	

Note 1: If the cable is installed as per installation methods 101 or 103 a larger csa cable is required or thermosetting cable per 4E3A.

Note 2: RCD (or RCBO) required.

Note 3: NP means not permitted.

Table C 9.4 Standard domestic final circuit lengths cartridge fuses BS 1361.

Cartridge fuses BS 1361							
Type of circuit	Cable csa pvc/pvc (mm²)	Fuse rating (A)	Maximum length for TN-C-S PME (m)		Maximum length for TN-S (m)		Installation method
			MCB	RCD or RCBO	MCB	RCD or RCBO	
Ring 13 A socket outlet	2.5/1.5	30	note 2	111	note 2	111	A, C, 100, 102, see note 1
Radial 13 A socket outlets	2.5/1.5	20	40	40	40	40	
Cooker (oven and hob)	6/2.5	30	53	53	28	53	
Oven (no hob)	2.5/1.5	15	32	32	32	32	
Immersion heater	2.5/1.5	15	32	32	32	32	
Shower to 30 A (7.2 kW)	6/2.5	30	note 2	53	note 2	53	
Shower to 40 A (9.6 kW)	10/4	45	note 2	66	note 2	66	
Storage radiator	2.5/1.5	15	30	30	30	30	
Fixed lighting (excl. switch drops)	1.5/1.0	5	53	53	53	53	All

Note 1: If the cable is installed as per installation methods 101 or 103 a larger csa cable is required or thermosetting cable per 4E3A.

Note 2: RCD (or RCBO) required.

Note 3: NP means not permitted.

Table C 9.5 Standard domestic final circuit lengths rewirable fuses (semi-enclosed) BS 3036.

Rewirable fuses (semi-enclosed) BS 3036							
Type of circuit	Cable csa pvc/pvc (mm²)	Fuse rating (A)	Maximum length for TN-C-S PME (m)		Maximum length for TN-S (m)		Installation method
			MCB	RCD or RCBO	MCB	RCD or RCBO	
Ring 13 A socket outlet; note 1	2.5/1.5	30	note 2	111	note 2	111	A, C, 100, 102; see note 1
Radial 13 A socket outlets	2.5/1.5 4.0/1.5	20 20	NP 68	NP 68	NP 68	NP 68	C
Cooker (oven and hob)	6/2.5	30	53	53	53	53	
Oven (no hob)	1.5/1.0	15	30	30	30	30	All
Immersion heater	1.5/1.0	15	30	30	30	30	
Shower to 30 A (7.2 kW)	6/2.5	30	note 2	55	note 2	53	A, C, 100, 102; see note 1
Shower to 40 A (9.6 kW)	10/4	45	66	66	66	66	
Storage radiator	2.5/1.5	15	51	51	51	51	
Fixed lighting (excl. switch drops); note 2	1.0/1.0 1.5/1.0	5 5	35 53	35 53	35 53	35 53	All

Note 1: If the cable is installed as per installation methods 101 or 103 a larger csa cable is required or thermosetting cable per 4E3A.

Note 2: RCD (or RCBO) required.

Note 3: NP means not permitted.

C 9.3   All-purpose standard final circuits

The following tables are based on a voltage drop of 5%. Circuits are suitable for the installation reference methods listed in the tables. Voltage drop limitations are calculated for the most onerous installation reference method listed for the cable/ overcurrent device combination. The load assumed is the rating of the overcurrent device.

Some values of maximum length in the tables have the notation(s), indicating that the length is limited by earth fault loop impedance. If you have an external loop impedance lower that that used in the table then longer lengths are possible.

Table C 9.6 All-purpose 5 A to 16 A radial final circuits using insulated 70°C thermoplastic (pvc) insulated and sheathed flat cable having copper conductors.

Protective device		Cable size (mm²)	Permitted installation reference methods	Maximum length (m)	
Rating (A)	Type			$Z_s \leq 0.8\ \Omega$ TN-S	$Z_s \leq 0.35\ \Omega$ TN-C-S
5	BS 1361 BS 3036	1.0/1.0	103,101,A,100,102,C	56 56	56 56
5	BS 1361 BS 3036	1.5/1.0	103,101,A,100,102,C	88 88	88 88
6	BS EN 60269–2,BS 88–6 MCB and RCBO Type B MCB and RCBO Type C MCB and RCBO Type D	1.0/1.0	103,101,A,100,102,C	46 46 46 25 (s)	46 46 46 36 (s)
6	BS EN 60269–2,BS 88–6 MCB and RCBO Type B MCB and RCBO Type C MCB and RCBO Type D	1.5/1.0	103,101,A,100,102,C	72 72 72 30 (s)	72 72 72 43 (s)
10	BS EN 60269–2,BS 88–6 MCB and RCBO Type B MCB and RCBO Type C MCB and RCBO Type D	1.0/1.0	101,A,100,102,C	26 24 24 8	26 24 24 18
10	BS EN 60269–2,BS 88–6 MCB and RCBO Type B MCB and RCBO Type C MCB and RCBO Type D	1.5/1.0	103,101,A,100,102,C	39 39 39 9	39 39 39 22
15	BS 1361 BS 3036	1.0/1.0	C NP	17 NP (ol)	17 NP (ol)
15	BS 1361 BS 3036	1.5/1.0	100,102,C NP	26 NP (ol)	26 NP (ol)
15	BS 1361 BS 3036	2.5/1.5	101,A,100,102,C 100,102,C	43 45	43 45
15	BS 1361 BS 3036	4.0/1.5	103,101,A,100,102,C 101,A,100,102,C	72 75	72 75
16	BS EN 60269–2,BS 88–6 MCB and RCBO Type B MCB and RCBO Type C MCB and RCBO Type D	1.0/1.0	C C	16 16 14 (s) NP (s)	16 16 16 8 (s)
16	BS EN 60269–2,BS 88–6 MCB and RCBO Type B MCB and RCBO Type C MCB and RCBO Type D	1.5/1.0	100,102,C 100,102,C	24 24 17s NP (s)	24 24 24 10 (s)

(Continued.)

Table C 9.6 (*Continued.*)

Protective device		Cable size (mm²)	Permitted installation reference methods	Maximum length (m)	
Rating (A)	Type			$Z_s \leq 0.8\,\Omega$ TN-S	$Z_s \leq 0.35\,\Omega$ TN-C-S
16	BS EN 60269–2, BS 88–6 MCB and RCBO Type B MCB and RCBO Type C MCB and RCBO Type D	2.5/1.5	101,A,100,102,C 101,A,100,102,C	40 40 27s NP (s)	40 40 40 15 (s)
16	BS EN 60269–2, BS 88–6 MCB and RCBO Type B MCB and RCBO Type C MCB and RCBO Type D	4.0/1.5	103,101,A,100,102,C 103,101,A,100,102,C	66 66 31 (s) NP (s)	66 66 54 (s) 18 (s)

Key:
NP, not permitted.
ol, cable/device/load combination not allowed in any of the installation conditions.
s, limited by earth fault loop impedance Z_s.

Table C 9.7 All-purpose 20 A to 32 A radial final circuits using insulated 70°C thermoplastic (pvc) insulated and sheathed flat cable having copper conductors.

Protective device		Cable size (mm²)	Permitted installation reference methods	Maximum length (m)	
Rating (A)	Type			$Z_s \leq 0.8\,\Omega$ TN-S	$Z_s \leq 0.35\,\Omega$ TN-C-S
20	BS EN 60269–2, BS 88–6 BS 1361 BS 3036 MCB and RCBO Type B MCB and RCBO Type C MCB and RCBO Type D	2.5/1.5	A,100,102,C A,100,102,C NP A,100,102,C	31 31 NP (ol) 31 14 (s) NP (s)	31 31 NP (ol) 31 31 9 (s)
20	BS EN 60269–2, BS 88–6 BS 1361 BS 3036 MCB and RCBO Type B MCB and RCBO Type C MCB and RCBO Type D	4.0/1.5	101,A,100,102,C 101,A,100,102,C C 101,A,100,102,C	48 (s) 44 (s) 48s 53 17s NP (s)	53 53 57 53 39 (s) 11 (s)
20	BS EN 60269–2, BS 88–6 BS 1361 BS 3036 MCB and RCBO Type B MCB and RCBO Type C MCB and RCBO Type D	6.0/2.5	103,101,A,100,102,C 103,101,A,100,102,C	77 (s) 71 (s) 77 (s) 81 27 (s) NP (s)	81 81 85 81 63 (s) 17 (s)

(*Continued.*)

Table C 9.7 (*Continued.*)

Protective device		Cable size (mm²)	Permitted installation reference methods	Maximum length (m)	
Rating (A)	Type			$Z_s \leq 0.8\,\Omega$ TN-S	$Z_s \leq 0.35\,\Omega$ TN-C-S
25	BS EN 60269–2, BS 88–6	2.5/1.5	C	26	26
	MCB and RCBO Type B		C	26	26
	MCB and RCBO Type C			5 (s)	24 (s)
	MCB and RCBO Type D			NP (s)	4 (s)
25	BS EN 60269–2, BS 88–6	4.0/1.5	A,100,102,C	31 (s)	42
	MCB and RCBO Type B		A,100,102,C	42	42
	MCB and RCBO Type C			5 (s)	28 (s)
	MCB and RCBO Type D			NP (s)	4 (s)
25	BS EN 60269–2, BS 88–6	6.0/2.5	101, A,100,102,C	50 (s)	64
	MCB and RCBO Type B		101, A,100,102,C	64	64
	MCB and RCBO Type C			9 (s)	45 (s)
	MCB and RCBO Type D			NP (s)	8 (s)
30	BS 1361	4.0/1.5	C	17s	36
	BS 3036		NP	NP (ol)	NP (ol)
30	BS 1361	6.0/2.5	A,100,102,C	27 (s)	53
	BS 3036		C	23 (s)	57
30	BS 1361	10/4.0	101, A,100,102,C	45 (s)	90
	BS 3036		A,100,102,C	37 (s)	93
32	BS EN 60269–2, BS 88–6	4.0/1.5	C	11 (s)	33
	MCB and RCBO Type B		C	31 (s)	33
	MCB and RCBO Type C			NP (s)	18 (s)
	MCB and RCBO Type D			NP (s)	NP (s)
32	BS EN 60269–2, BS 88–6	6.0/2.5	A,100,102,C	19 (s)	49
	MCB and RCBO Type B		A,100,102,C	49	49
	MCB and RCBO Type C			NP (s)	29 (s)
	MCB and RCBO Type D			NP (s)	1 (s)

Key:

NP, not permitted.

ol, cable/device/load combination not allowed in any of the installation conditions.

s, limited by earth fault loop impedance Z_s.

Table C 9.8 Ring final circuits using insulated 70°C thermoplastic (pvc) insulated and sheathed flat cable having copper conductors.

Protective device		Cable size (mm²)	Permitted installation reference methods	Maximum length m	
Rating (A)	Type			$Z_s \leq 0.8\,\Omega$ TN-S	$Z_s \leq 0.35\,\Omega$ TN-C-S
30	BS 1361 BS 3036	2.5/1.5	A,100,102,C	59 (s) 49 (s)	111 111
30	BS 1361 BS 3036	4.0/1.5	101,A,100,102,C	183 69	183 159
32	BSEN 60269–2,BS88–6 MCB and RCBO Type B MCB and RCBO Type C MCB and RCBO Type D	2.5/1.5	A,100,102,C	106 106 NP (s) NP (s)	106 106 63 1
32	BSEN 60269–2,BS88–6 MCB and RCBO Type B MCB and RCBO Type C MCB and RCBO Type D	4.0/1.5	101,A,100,102,C	176 176 NP (s) NP (s)	176 176 33 4

Key:

NP, not permitted.

ol, cable/device/load combination not allowed in any of the installation conditions.

s, limited by earth fault loop impedance Z_s.

 ## C 10 RCDs and circuitry

C 10.1 Introduction, increased use of RCDs

Chapter D discusses the selection of RCDs and their increased specification in BS 7671: 2008 compared with previous editions. The increased use of RCDs includes the following:

● All socket-outlet circuits accessible to ordinary or non-instructed persons* (411.3.3).
● All circuits within a bathroom (701.411.3.3).
● Concealed cables either less than 50 mm deep, in installations not intended to be under the supervision of a skilled or instructed person*, not mechanically protected or protected by earthed metal and not run in the 'safe zones' *.
● Concealed cables within metal partitions, in installations not intended to be under the supervision of a skilled or instructed person*, not mechanically protected or protected by earthed metal and accessible to ordinary or non-instructed persons*.

- Generally where disconnection times cannot be achieved with an overcurrent device.
- Most TT installations, usually required to achieve a disconnection time of 0.2 s.
- Other special locations, including swimming pools and caravan parks.

*The subject of providing RCDs for ordinary persons is covered in D 7.1, which should be read in conjunction with this section. They are now required for most domestic socket-outlet circuits as well as some commercial and industrial socket-outlet circuits, and this section of the book details how to arrange the circuitry aspects of RCDs in order to maintain compliance with the Regulations.

C 10.2   Consumer unit arrangements for RCDs

Every installation shall be divided into circuits as necessary to:

- avoid hazards and minimize inconvenience in the event of a fault;
- facilitate safe inspection, testing and maintenance;
- take account of danger that may arise from circuit failure such as a lighting circuit;
- reduce the possibility of unwanted tripping of RCDs, due to excessive protective conductor currents produced by equipment in normal operation;
- mitigate the effects of electromagnetic interference; and
- prevent the indirect energizing of a circuit intended to be isolated.

There are various ways of arranging RCDs in consumer units; an obvious but expensive option is to use RCBOs for every circuit. Short of this, the figures in this section show some useful arrangements.

Figure C 10.1 shows a split load consumer unit with single 30 mA RCD. This would be used in both TN-S and TN-C-S systems and provide compliance with the 'Additional Protection' requirements of Chapter 41.

Figure C 10.2 shows an installation wholly protected with RCDs, a 100 mA main switch RCD incorporating a time delay and a secondary 30 mA RCD. This is suitable for TN-S and TN-C-S installations, and TT installations (with an insulated consumer unit).

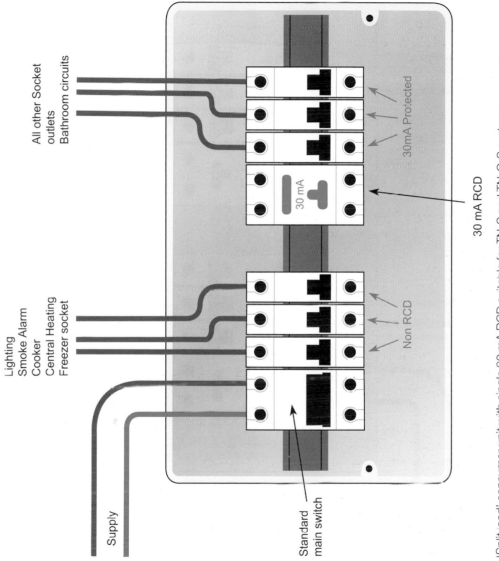

All other Socket outlets
Bathroom circuits

30mA Protected

30 mA

30 mA RCD

Lighting
Smoke Alarm
Cooker
Central Heating
Freezer socket

Non RCD

Supply

Standard
main switch

Figure C 10.1 'Split load' consumer unit with single 30 mA RCD suitable for TN-S and TN-C-S systems.

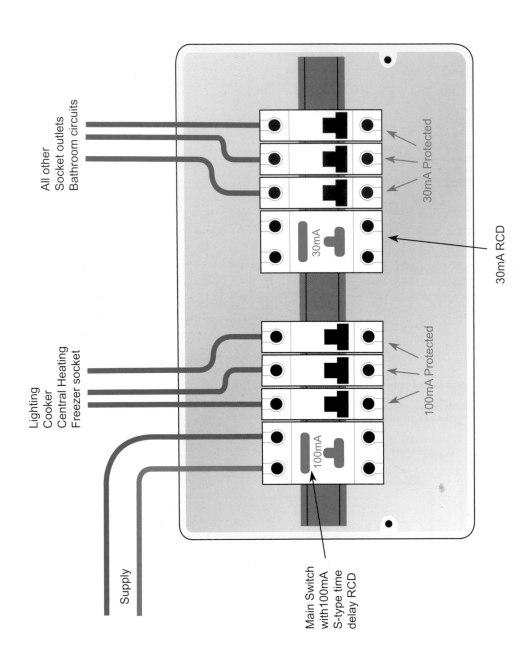

All other
Socket outlets
Bathroom circuits

30mA Protected

30mA

Lighting
Cooker
Central Heating
Freezer socket

100mA Protected

100mA

30mA RCD

Supply

Main Switch
with100mA
S-type time
delay RCD

Figure C 10.2 Consumer unit with main switch time-delayed RCD and secondary RCD suitable for TN-S, TN-C-S and TT systems.

C 11 **Ring and radial final circuits**

C 11.1 Introduction

BS 7671: 2008 provides some basic guidance on ring and radial final circuits. This information is included within Appendix 15 of BS 7671: 2008, the diagrams of which are reproduced in this section with kind permission of the IET.

C 11.2 Ring final circuits

Rings must comply with Regulation 433.1.4, which is as follows:

> **433.1.4** Accessories to BS 1363 may be supplied through a ring final circuit, with or without unfused spurs, protected by a 30 A or 32 A protective device complying with BS 88–2.2, BS 88–6, BS 1361, BS 3036, BS EN 60898, BS EN 60947–2 or BS EN 61009–1 (RCBO). The circuit shall be wired with copper conductors having line and neutral conductors with a minimum cross-sectional area of 2.5 mm² except for two-core mineral insulated cables complying with BS EN 60702–1, for which the minimum cross-sectional area is 1.5 mm². Such circuits are deemed to meet the requirements of Regulation 433.1.1 if the current-carrying capacity (I_z) of the cable is not less than 20 A and if, under the intended conditions of use, the load current in any part of the circuit is unlikely to exceed for long periods the current-carrying capacity (I_z) of the cable.

Recommendations on how to achieve this are as follows, noting that this advice may differ from that in Appendix 15 of BS767: 2008:

● For commercial and industrial installations radials are preferred.
● Large fixed loads (2 kW and above) should not be connected near the 'ends' of a 32 A ring.
● Kitchen appliances should preferably be supplied via dedicated radials or rings.

Figure C 11.1 is based upon that in Appendix 15 of BS 7671: 2008.

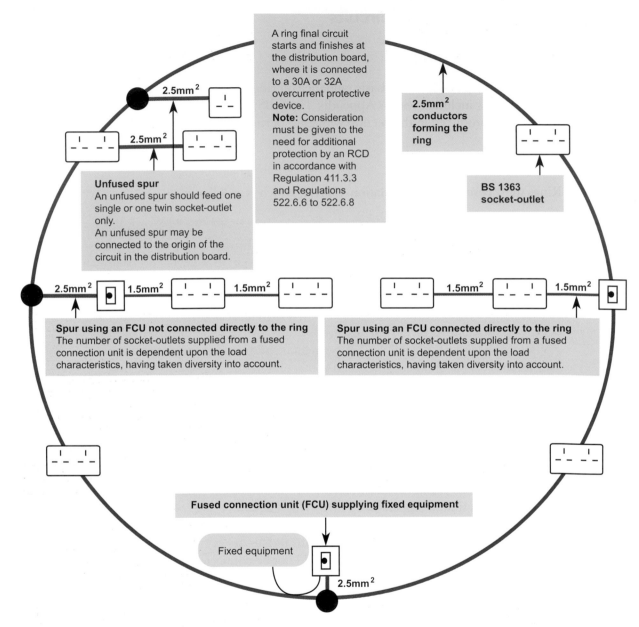

A ring final circuit starts and finishes at the distribution board, where it is connected to a 30A or 32A overcurrent protective device.
Note: Consideration must be given to the need for additional protection by an RCD in accordance with Regulation 411.3.3 and Regulations 522.6.6 to 522.6.8

2.5mm²

2.5mm²

2.5mm² conductors forming the ring

BS 1363 socket-outlet

Unfused spur
An unfused spur should feed one single or one twin socket-outlet only.
An unfused spur may be connected to the origin of the circuit in the distribution board.

2.5mm²   1.5mm²   1.5mm²

1.5mm²   1.5mm²

Spur using an FCU not connected directly to the ring
The number of socket-outlets supplied from a fused connection unit is dependent upon the load characteristics, having taken diversity into account.

Spur using an FCU connected directly to the ring
The number of socket-outlets supplied from a fused connection unit is dependent upon the load characteristics, having taken diversity into account.

Fused connection unit (FCU) supplying fixed equipment

Fixed equipment

2.5mm²

Figure C 11.1 Ring final circuit information diagram.

C 11.3 Radial final circuits

Figure C 11.2 is based upon that in Appendix 15 of BS 7671:2008.

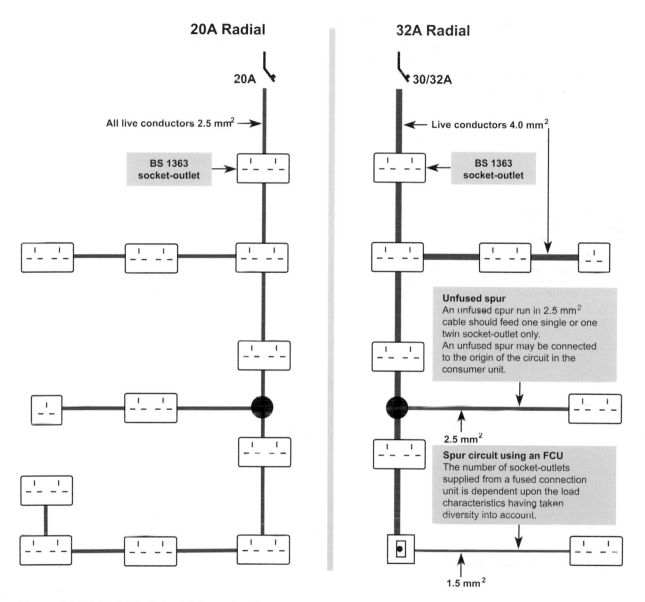

Figure C 11.2 Radial final circuit information diagram.

Selection and Erection – Equipment

D 1 Introduction and fundamentals

This chapter on selection and erection of equipment discusses and provides guidance and solutions on what essentially is an equipment area of BS 7671: 2008, and includes a sizeable amount of the Regulations. Consistent with the general philosophy of this book, all areas are amply covered with particular practical guidance on more difficult or contentious areas.

The selection of equipment generally is a very important aspect for the designer, but often understated is that the installer selects equipment. In this case the installer is carrying out part-design, and installers should obtain selection advice from their designer unless contracted to do otherwise.

The term *equipment*, as the BS 7671 definition makes clear, relates to all equipment to be utilized in an electrical installation. So as to be clear as to what is included, the BS 7671 definition of *electrical equipment* is repeated in the box below.

> *Electrical equipment (abbr: equipment). Any item for such purposes as generation, conversion, transmission, distribution or utilization of electrical energy, such as machines, transformers, apparatus, measuring instruments, protective devices, wiring systems, accessories, appliances and luminaires.*

Fundamental requirements for selecting all equipment

It is important to note that the selection and erection requirements of the 17th Edition do not in general repeat requirements within the product standards to which the equipment is made. Instead, BS 7671 addresses equipment so far as selection and equipment within the installation is concerned (Regulation 113.1).

When selecting equipment, the following must be considered:

- compliance with the appropriate product standards;
- suitability for the anticipated operational conditions;
- suitability for the anticipated external influences;
- provision for adequate accessibility for maintenance.

D 2   Compliance with Standards

One of the most fundamental rules of the 17th Edition is that equipment shall comply with an acceptable and current equipment Standard. Most equipment will have a specific British Standard or BS EN written for it, and these standards would need to be used for specifying or selecting equipment.

However, for many reasons, it may not be possible or desirable to select equipment to an appropriate BS or BS EN Standard, and in these cases alternatives are permissible. In the author's experience this route is not uncommon as often non-European equipment is more viable; an example here is the prefabricated wiring systems made to IEC or USA Standards. In cases such as this there are two routes to compliance and the flow diagram below provides some further guidance.

Firstly, equipment to a foreign national Standard which is based on the corresponding IEC Standard may be used provided that, as the 17th Edition requires: 'the designer confirms that it provides at least the same degree of safety as equivalent British Standard equipment'. In practice this exercise will usually approach the impossible and, in many cases, a sensible view has to be taken. Some will find this judgement to be unacceptable but it is your responsibility, if you are the designer; you may wish to use only BS EN equipment.

Another route applies where you wish to use equipment not covered by a Standard, or where it is used outside of scope, and here the designer must confirm its safety to the 17th Edition. This option is at least achievable.

A diagram of how to select a product standard in relation to Section 511 is given in Figure D 2.1.

It should now be obvious that selection of equipment is very much dependent upon knowledge of the equipment Standards and, from this aspect, the BSI website is invaluable.

This book includes a list of some relevant BS and BS EN standards in Appendix 1.

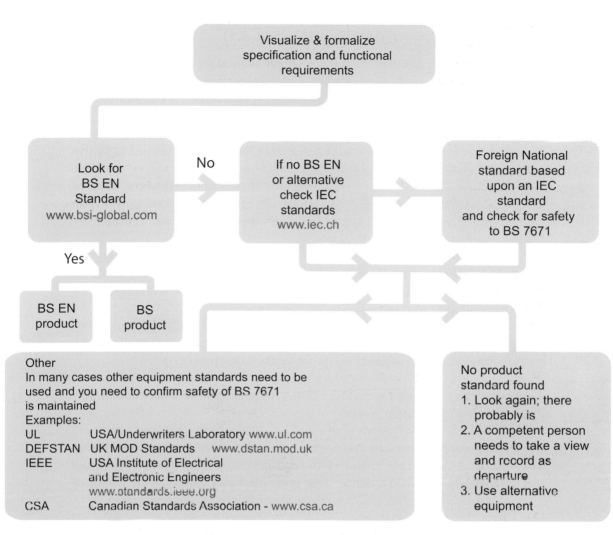

Figure D 2.1 Selection of equipment – Standards flow diagram.

D 3 Identification of conductors

Introduction

Harmonized cable colours were introduced into BS 7671 in the 2004 Amendment, and transitional arrangements were made. The 17th Edition makes no changes to the regulations on cable identification or to Appendix 7 of BS 7671, which provides details of interfaces with existing installations.

D 3.1 Principle of required identification (514.3.1)

It is most important to understand the principle behind the drafting of the 17th Edition in respect of cable identification. Cable cores shall be identifiable at their terminations either by colour or by alphanumeric characters. While this has not changed from the 16th Edition it is worth discussing the principles.

It is noted that cores should preferably be identifiable throughout their length. For many applications coloured cables, either single- or multicore, will be used and these cables are obviously identified throughout their length. However, in many other applications installers will need to make use of this Regulation (514.3.1) for overmarking at terminations. The principles and applications of identification are now discussed with the use of diagrams.

Figure D 3.1 shows the principle of identification where colours are used, i.e. marking by colour at all terminations and preferably throughout the length.

In Figure D 3.1 marking throughout the length is not used and single core cables have been used. It should be noted that the colour of the cables originally used is not important and overmarking by tape or similar takes precedence.

Building on this principle, Figure D 3.2 shows the principle of identification where alphanumeric identification is used.

It is perhaps more obvious now that marking throughout the length is not necessary. It should be noted that the colour of the cables originally used is not important and overmarking takes precedence. This principle holds whatever the colour of cores of the original cable, and applies at terminations by coloured tapes or by characters.

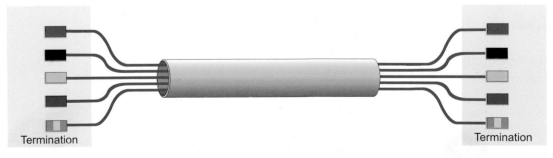

Termination   Termination

Figure D 3.1 Principle of colour identification.

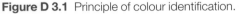

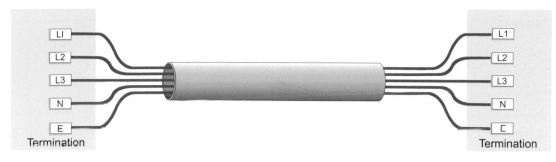

Figure D 3.2 Principle of identification by characters.

In summary, an important principle is established in that, wherever marking at terminations by either tapes, lettering or numbering is used, the original cable colours must be ignored; this can be against your gut feeling but you must get used to it!

Common examples and applications where identification is only practised at terminations include the following:

- Multicore cables with more than five cores.
- MICC cables.
- Control applications wired in single-core conductors.
- Applications where coloured cable is not available, including connections to large generators or transformers, where often only black cable is available.

It should now be established that installers may wish to use any single cable colour, or combination of colours, and overmark at terminations, and this is not considered a lesser option. For green-and-yellow conductors in multicore cables, overmarking in another colour at terminations is permitted. Overmarking at terminations is prohibited for single-core green-and-yellow conductors.

The general principles are now discussed further and some applications are included.

D 3.2   Identification by colour

Where colour is used to comply with the method of identification, Table 51 from BS 7671: 2008 gives the colour options.

Table 51 Identification of conductors (from BS 7671: 2008).

Function	Alphanumeric	Colour
Protective conductor		Green-and-yellow
Functional earthing conductor		Cream
a.c. power circuit [3]		
Line of single-phase circuit	L	Brown
Neutral of single- or three-phase circuit	N	Blue
Line 1 of three-phase circuit	L1	Brown
Line 2 of three-phase circuit	L2	Black
Line 3 of three-phase circuit	L3	Grey
Two-wire unearthed d.c. power circuit		
Positive of two-wire circuit	L+	Brown
Negative of two-wire circuit	L−	Grey
Two-wire earthed d.c. power circuit		
Positive (of negative earthed) circuit	L+	Brown
Negative (of negative earthed) circuit [4]	M	Blue
Positive (of positive earthed) circuit [4]	M	Blue
Negative (of positive earthed) circuit	L−	Grey
Three-wire d.c. power circuit		
Outer positive of two-wire circuit derived from three-wire system	L+	Brown
Outer negative of two-wire circuit derived from three-wire system	L	Grey
Positive of three-wire circuit	L+	Brown
Mid-wire of three-wire circuit [1,4]	M	Blue
Negative of three-wire circuit	L−	Grey
Control circuits, ELV and other applications		
Line conductor	L	Brown, Black, Red, Orange, Yellow, Violet, Grey, White, Pink, Turquoise
Neutral or mid-wire [2]	N or M	Blue

[1] Only the middle wire of three-wire circuits may be earthed.

[2] An earthed PELV conductor is blue.

[3] Power circuits include lighting circuits.

[4] M identifies either the middle wire of a three-wire d.c. circuit, or the earthed conductor of a two-wire earthed d.c. circuit.

Following the principles of the Regulations, it should be stressed that the alphanumeric column was added to Table 51 for reference and if the colour option is used, these are not required.

Control circuits, ELV and other applications

It should be noted that in the last part of Table 51 'Control circuits, ELV and other applications' the use of other cable colours is allowed for these particular applications. This can be for any application where the general standard power colours are not preferred. An example of where this would be used would be a lock-stop button control cable, where it is desired to use, say, violet as a colour. It should be noted that additional termination marking (by colour or characters) would not be necessary (the information relating to violet as a colour for the lock-stop cable would normally be contained on schedules or schematic diagrams).

D 3.3   Identification by marking

Colour identification does not have to be used, and of course sometimes cannot be used, for example when using MICC cables. In these cases, or where the designer or installer simply does not prefer colour, identification by marking at terminations is used. Section 514 of BS 7671: 2008 allows you the flexibility to select an identification system that works for you and the application.

This may either be the alphanumeric identification of Table 51 or can be by numbering (514.5.4). Numbering may be the preferred method of identification; for example, multicore armoured termination kits. Take a 27-core armoured cable. The cores are numbered from 0 to 26; in the installation, a number of these may be used for neutrals and others used for earths (514.5.4 states that the number 0 should be used for the neutral). They are all coloured black. Normally, to satisfy other parts of BS 7671, a suitable wiring diagram or equivalent will be required, indicating the use of the numbers (unless there is no possibility of confusion).

D 3.4   Additions and alterations – identification

Where pre-2004 harmonized cable colours (generally red, yellow, blue and black) and harmonized colours (generally brown, black, grey and blue) are used in the same installation, the following label must be applied at the relevant distribution boards or items of switchgear as appropriate.

> **CAUTION:** *This installation has wiring colours to two versions of BS 7671. Great care should be taken before undertaking extension, alteration or repair that all conductors are correctly identified.*

It should be noted that this label need not be applied at all other points on circuits, only at the point of the circuits most likely to be used for isolation in the future, most commonly the distribution boards or items of switchgear.

D 3.5   Interface marking

When harmonized cable colours were introduced into BS 7671 in the 2004 Amendment, a new Appendix 7 was included detailing best practice for marking between pre-harmonized and harmonized colours. This appendix is reproduced in the 17th Edition.

The 17th Edition requires that termination overmarking is applied 'except where there is no possibility of confusion' at interfaces (the interface is the joint between pre-harmonized and harmonized).

In practice, the installer will need to learn the principle and apply it accordingly, but no marking is required for most single-phase correctly coloured interfaces – see Figure D 3.3.

For three-phase applications, Appendix 7 of BS 7671 suggests applying L1, L2, L3 and N at interfaces, again, where there is no possibility of confusion. The requirements are illustrated for some typical three-phase installation interfaces.

There are examples when interface marking will not be required, but the decision is for the installer, and if you are uncomfortable with interpretation it is recommended that you always apply the interface marking. However, there are of course applications where interface marking is not required, as per the Figure D 3.4.

D 3.6   d.c. identification

d.c. colour or alphanumeric identification is given in Table 51 within the 17th Edition, and both the colours and alphanumeric specified have caused some heartache since they were introduced in the 2004 Amendment. Many engineers feel that the d.c. colours specified in the 17th Edition are inappropriate; this is a classic area where, if it is felt that existing practice of colours is less confusing, a departure is made with an appropriate note recorded on the completion certificate.

	General Warning Label Required	Interface Marking Required
	At Switchgear	Not Required
	At Switchgear	Mark browns with L1, L2, L3 & Neutral conductor with N at load and preferably at switchgear
	At Switchgear	Not Required
	At Switchgear	Not Required

New L1 load

New three phase load

New L2 load

New L3 load

NOTES: All single core circuits like these 'tap-offs' shall be in Brown phase conductors (applies to main, sub-main & final circuits), where colour is used.

Existing Switchgear e.g. Busbar

Figure D 3.3 Typical extension interface marking.

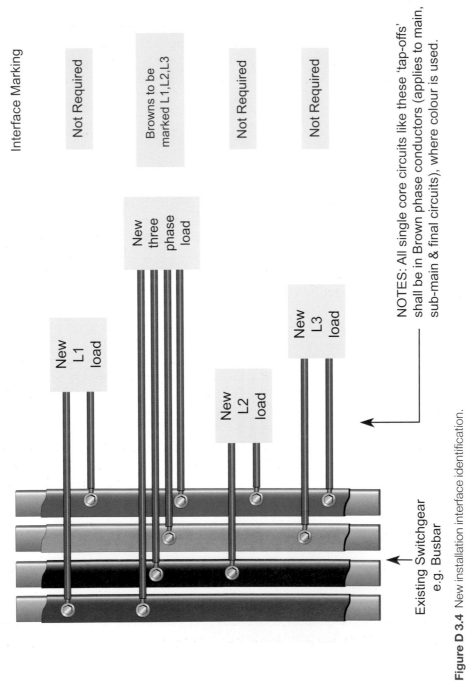

Interface Marking

Not Required

Browns to be marked L1,L2,L3

Not Required

Not Required

New L1 load

New three phase load

New L2 load

New L3 load

NOTES: All single core circuits like these 'tap-offs' shall be in Brown phase conductors (applies to main, sub-main & final circuits), where colour is used.

Existing Switchgear e.g. Busbar

Figure D 3.4 New installation interface identification.

D 4 EMC and prevention of mutual detrimental influences

D 4.1 Introduction

The Wiring Regulations have for a number of years included basic regulations for electromagnetic compatibility (EMC), regulations about separating system voltages and also regulations about the rather cumbersome title of 'preventing mutual detrimental influences'. These topics are all discussed in this section as many of the recommendations overlap.

BS 7671: 2008 has a number of regulations to take note of as follows:

● Section 528 – proximity to other cables, equipment.
● Part 515 EMC.

Let's first look at proximity; Regulation 528.2 requires that ELV and LV circuits shall not be in the same containment unless insulated for the highest voltage present. This applies to cables on cable tray as well as within conduit, trunking and ducting. Of course, this insulation requirement is not required where compartments physically separate the cables.

D 4.2 EMC directive and BS 7671

Previously, the EMC directive did not specifically mention installations, but the recent revision to the directive now states installations are not exempt from the legislation. This section provides brief guidance on complying with the EMC directive and EMC requirements of BS 7671: 2008.

The directive and official guidance indicates that best practice is followed, and cites HD 60384, the European harmonization document upon which BS 7671: 2008 is based. Regulation 515.3 states that equipment shall be chosen with suitable immunity and emission levels.

The EMC directive Regulations and various guidance documents can be found on the DTI website, and this also provides links for supporting informative EMC documents as well as a link to the EU Commission website. All main documents and guides on these sites were, at the time of writing, free of charge.

So, what is the impact and what advice can be provided to installation designers and installers? Three key points of action are required:

1   Equipment should preferably be to BS EN standards. Immunity and emission are written into standards and, irrespective of this, manufacturers have an obligation to manufacture to the EU EMC levels. If you need to use equipment to a different standard it is a good idea to obtain confirmation that equivalent immunity and emission has been achieved. It is important to note that it is the EU retailer who has the duty to confirm this (e.g. the importer). This is not always as difficult as it sounds, as Standards like the USA's FCC rules are both stringent and well documented.

2   Design and install circuits and equipment to 'best practice'. This means to BS 7671 and other recommended guides. At the time of writing, a new section (444) was being drafted by the Wiring Regulations committee and is likely to include new regulations and some guidance on this subject.

3   Formalize documentation. It is perhaps even more important now to formalize the storage of the project's technical design and installation information.

By completing these three actions, designers and installers will have fulfilled their duties under the EMC directive as well as BS 7671: 2008.

D 4.3   EMC cable separation – power, IT, data and control cables

General

Firstly it must be stated that the majority of power installations installed to BS 7671 will exhibit no EMC problems between the power system and 'sensitive' data type installation. Indeed, it is interesting to note that many data IT installation EMC problems have been discovered to be infringements of BS 7671, particularly poor earthing.

In spite of this, the following guidance is recommended where 'sensitive' equipment is installed ('sensitive' equipment means IT data and communications equipment). Some of the recommendations will have to be balanced against cost and performance.

When the requirements for the segregation of sensitive equipment and cables are being considered for an installation, more design and installation information can be found in the series of Standards BS EN 50310, BS EN 50173 and BS EN 50174 Part 1–3.

Recommendations – cable separation for sensitive equipment

The following optional solutions may be considered to mitigate the effects of EMC on sensitive equipment:

- separate power transformers as Figure D 4.1;
- size neutral for unbalanced load and triple harmonic currents[1];
- balance loads;
- in extreme cases, oversizing the power transformer to lower the source impedance.

[1] The subject of harmonic assessments is covered in Chapter C.

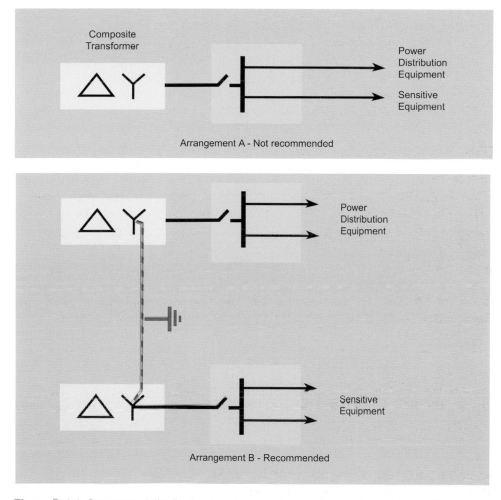

Figure D 4.1 Segregated distribution transformers for sensitive equipment installations.

The following points are also recommended to prevent the electricity distribution systems adversely influencing sensitive equipment. There will be practical limits to most of the options, and compromises will often be necessary:

1   A physical distance-based separation between potential sources of interference, e.g. lifts, power transformers, variable speed drives, high current busbars or HV equipment.

2   Metal pipes such as water, gas, heating and cables should enter the building at the same location as the electrical power cable and be connected to the main equipotential bonding system at this point.

3   A common route through the building for power and signal cables will avoid large inductive loops formed by different cabling systems.

4   The necessary separation between power and information technology cables by distance or by screening should be provided. Separation distances are found in Table D 4.1.

5   Power and information technology cables should cross over at right angles.

6   Power cabling systems that use single-core conductors should ideally be enclosed in earthed metallic enclosures or conduits.

The separation distances given in Table D 4.1 are for backbone cabling from end to end. However, for horizontal cabling where the final circuit length is less than 35 metres no separation is required in the case of screened cabling.

Table D 4.1 Recommended cable separation distances.

Type of installation	Minimum separation distance		
	Without divider or non-metallic divider[1]	Aluminium divider	Steel divider
Unscreened power cable and unscreened IT cable	200 mm	100 mm	50 mm
Unscreened power cable and screened IT cable[2]	50 mm	20 mm	5 mm
Screened power cable and unscreened IT cable	30 mm	10 mm	2 mm
Screened power cable and screened IT cable[2]	0 mm	0 mm	Separation by distance or by screening

[1] It is assumed that in case of metallic divider, the design of the cable management system will achieve a screening attenuation related to the material used for the divider.
[2] The screened IT cables shall comply with EN 50288 series.

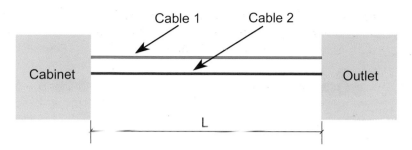

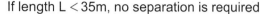

If length L < 35m, no separation is required

Figure D 4.2 Schematic of length and required separation recommendations.

For lengths greater than 35 metres, the separation shall apply to the full length, excluding the final 15 metres before it is attached to the outlet (see Figure D 4.2).

D 4.4 Cable management and EMC

Some metallic cable management products offer an improved protection from electromagnetic emissions (EMI). Plastic cable management systems can also offer some protection but are only recommended where the system has a low emission level, or where optical fibre cabling is used.

Figure D 4.3 illustrates recommended cable layouts within cable management products.

Where metallic system components are used, the shape (flat, U-shape, tube, etc.) will determine the characteristic impedance of the cable management system. Enclosed shapes, by reducing the common mode coupling, offer the best profile.

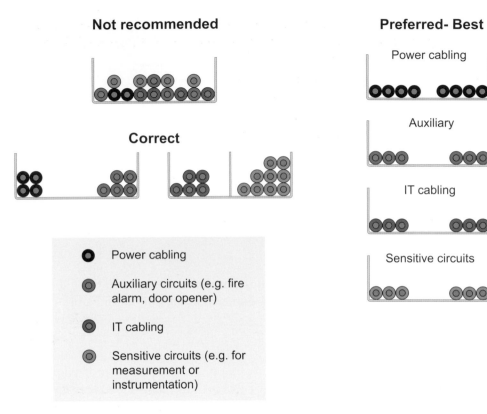

Figure D 4.3 Cable separation within cable management systems.

D 5 Wiring systems

D 5.1 The choice of wiring systems

The choice of wiring system will depend on a number of factors that the installation designer will need to consider, depending on the particular type of installation, its use, the environment and also any economic constraints.

In order to achieve compliance with BS 7671, it is important to select a type of wiring system complying with the appropriate product standard, as previously mentioned in Section D 2.

The selection of an appropriate wiring system can be represented as a balancing act between specification, cost and time, as depicted in the balance triangle below:

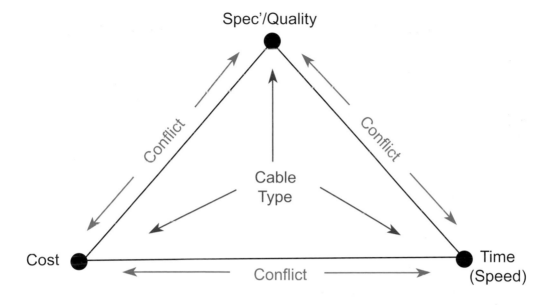

The balance of any one of the three selection parameters is in conflict with the other two, and in selecting any one cable type, compromises will need to be made. Maintenance considerations are not shown on this diagram but should be considered under the specification parameter. The complex criteria of specification will be very much affected by environmental factors and the use by operators. With this in mind, the following points will need to be considered in the specification parameter:

- type of installation, domestic/commercial/industrial and impact protection;
- temperature of installation and local heat sources;
- effects of dust and water;
- effects of chemicals, fumes and gases;
- animals including vermin;
- movement and mechanical vibrations;
- corrosion including electrolytic corrosion;
- other environmental factors including wind, seismic effects, solar radiation, hygiene and mould growth.

This list is not exhaustive, and experience and knowledge are key factors in selecting the wiring system.

A key 17th Edition regulation in cable selection is the mechanical protection, or impact protection, as required under the 522.6 group of regulations.

A concealed cable installed within 50 mm of the surface of a building fabric (shallow) needs careful consideration as listed in 522.6.5, 522.6.6, 522.6.7 and 522.6.8. The principle of these regulations is as follows:

● the running of concealed cables in zones 'deemed to be safe'; or,
● mechanical protection or 'earthed metallic' protection.

The permitted cable zones are indicated in Figure D 5.1.

Two additional regulations have been added in BS 7671: 2008 to the 522.6 'concealed cable protection' group of regulations as follows:

● 522.6.7 requires a 30 mA RCD for cables less than 50 mm deep run in the 'safe zones' where the installation is not intended to be under the supervision of a skilled or instructed person, and where the cable is not mechanically protected or not within earthed metallic enclosures.
● 522.6.8 requires a 30 mA RCD for cables of any depth concealed within metal partitions where the installation is not under the supervision of a skilled or instructed person, and is not mechanically protected, run in zones or within earthed metallic enclosures.

Generally, these applications 'not under the supervision of a skilled or instructed person' include domestic installations and parts of an installation where members of the public may use socket outlets (e.g. internet cafés). To save repetition, for interpretation on where to use RCDs and the extent to which instructed or skilled persons are included, see Section D 7.1.

To comply with 522.6 it should be noted that not all cables have to be installed in 'heavy-duty conduit'. As an example, cables to BS 8436, which are soft or semi-rigid cables incorporating an aluminium screen, can provide a solution. This relatively new cable has a screen that is in contact with a bare cpc, negating the need for special 'earthing' glands, and may be a solution in some cases.

Where unsheathed cables are to be used, the 'earthed metallic' enclosures include:

● steel conduit;
● flexible steel conduit;
● steel trunking;
● steel underfloor trunking.

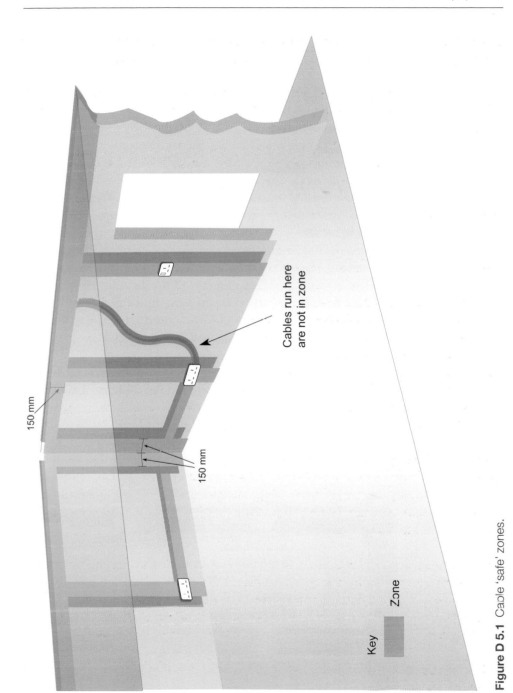

150 mm

150 mm

Cables run here
are not in zone

Key

Zone

Figure D 5.1 Cable 'safe' zones.

D

D 5.2 Circulating currents and eddy currents in single-core installations

Regulations 521.5.2 and 523.10 give requirements concerning ferrous enclosures and single-core armoured cables with respect to eddy currents and circulating currents. While 521.5.2 stipulates that steel wire armoured cables shall not be used with a.c. circuits there are other pitfalls in selecting alternative cables.

Single-conductor non-ferrous metal sheathed or armoured cables need careful consideration due to the problems of sheath currents and eddy currents.

Circulating (sheath) currents

Circulating or sheath currents flow in the metal sheaths or armour of single-core cables. All single-core metallic-sheathed cables have a sheath current induced by the a.c. magnetic field surrounding each conductor. The metallic sheaths and armoured cables are often earth-connected to equipment at both ends, providing a closed circuit path through metal gland plates, enclosures and other metallic paths where sheath currents are permitted to flow, as shown in Figure D 5.2.

Induced sheath currents cause a heating effect and temperature rise in metal sheaths or armour of single-conductor cables. This is transferred to the cable insulation and,

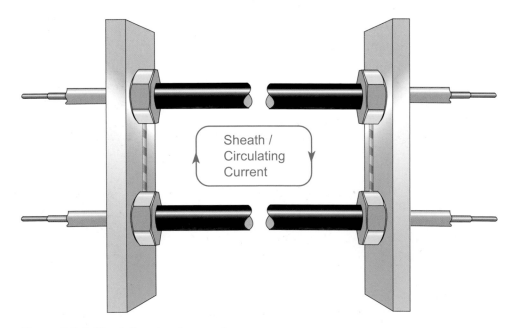

Figure D 5.2 Circulating sheath current.

depending on the general circuit loading, this additional heat may cause the cable to suffer premature insulation damage.

There is no recognized method for an exact calculation of the heating effect of sheath currents and the following options are advised:

Mitigating sheath (circulating) currents – option A

De-rate the single-conductor cable current rating by a factor of up to 50%. This will have cost and termination constraints, i.e. you may simply not be able to terminate the cable. These cables are often transformer or generator final connections and sizing should be rechecked; it is a fact that in many installations power cables never achieve anything approaching their operating temperature due to overdesign. There are many installations where single core cables are bonded (earthed) at both ends and problems virtually never exist. This is particularly true of the main cables between transformer and switchgear, which in many installations only ever reach a small fraction of their rated current. Whilst this advice may seem perhaps a bit open, ultimately an engineering judgement is required.

Mitigating sheath (circulating) currents – option B

A second option is providing isolation of the metallic sheath or armour from earth at one end of the cable. This is often achieved by cable termination with a non-metallic gland plate. This is sometimes known as single point bonding. Separate earth cables will be required to maintain the system earthing. To ensure that metallic sheaths or armour are isolated from earthed metal, cables will almost certainly need a non-metallic outer sheath covering. This will ensure that sheath currents will not flow spuriously due to contact with metallic parts such as cable trays, racks or structural steel. Due to the possibility that isolation is not always maintained, option A might be preferred. If this method is being used, it is recommended to limit the longitudinal voltage to 25 volts for corrosion and safety reasons. This is estimated using the following rule of thumb:

For trefoil cables, voltage = 0.05 mV/A/m.

For flat formation, voltage = 0.125 mV/A/m.

Eddy currents

The same magnetic fields that surround single-conductor cables can also produce eddy currents in the steel enclosures, which completely surround the cables.

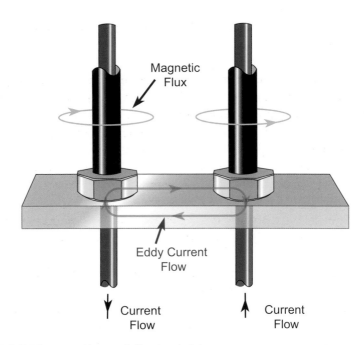

Figure D 5.3 Eddy current in a metallic gland plate.

Figure D 5.3 shows how and where eddy currents are created at a metallic gland plate.

Eddy currents can overheat iron or steel cabinets, locknuts or bushings or any ferrous metal that completely encircles the single-conductor cables. This presents no problem in multi-conductor cables, where the magnetic fields tend to cancel each other out. For single-core cables, it is recommended that these cables enter metal enclosures through a non-ferrous plate (e.g. aluminium). Any connectors, glands and similar that completely surround the conductors must be of non-ferrous materials. Slotting the equipment enclosure between the cable openings was considered an acceptable way to reduce the effects of eddy currents, but is no longer common practice (some feel that the practice of slotting weakens equipment).

D 5.3 Electrical connections and joints

D 5.3.1 General

A typical installation comprises a multitude of joints and connections, both in conductors and in wiring containment systems. As is seen from reading other chapters of this book, the soundness of connections is crucial, and loose or partially loose connections can cause all sorts of problems, including overheating and possibly fire.

It is imperative that such connections are made in such a way as to provide for durable electrical continuity and low resistance. It is important that connections in both live conductors (phase and neutral) exhibit adequate mechanical strength and continuity to allow the current-using equipment of the circuit concerned to function as intended. For protective conductors it is perhaps just as important that such connections are soundly made, because poor connections will remain undetected until the time of a possible earth fault.

The correct tightness of connections is essential to avoid particular risks from fire or other harmful thermal effects, such as:

● Excessive temperature developed in a high resistance connection when it carries current (e.g. load current or short-circuit current).
● Arcs or high temperature particles emitted from a loose connection.
● Poor connections giving rise to the risk of electric shock due to a malfunction of the protective measure of fault protection. A high resistance or open circuit in a protective conductor affecting the earth fault loop impedance may prevent automatic disconnection. Subsequent contact of a person or livestock with an exposed-conductive-part could be hazardous.

For the above considerations, BS 7671 makes a number of requirements concerning the construction of electrical connections and the workmanship necessary in carrying out such connections. All connections between conductors and between conductors and other equipment, such as current-using equipment, accessories or switchgear, are required to provide for durable electrical continuity and adequate mechanical strength (Regulation 526.1).

When selecting a suitable connection, BS 7671 requires the designer to take account of a number of factors (taken directly from Regulation 526.2), as appropriate:

● the material of the conductor and its insulation;
● the number and shape of the wires forming the conductor;
● the cross-sectional area of the conductor;
● the number of conductors to be connected together;
● the temperature attained at the terminals in normal service such that the effectiveness of the insulation of the conductors connected to them is not impaired;

- the provision of adequate locking arrangements in situations subject to vibration or thermal cycling;
- where a soldered connection is used the design shall take account of creep, mechanical stress and temperature rise under fault conditions.

Equipment may come with the following lettering on connection terminals:

- 'R' signifies that the terminal is suitable to accept only a rigid conductor.
- 'F' signifies that the terminal is suitable to accept only a flexible conductor.
- 'S' or 'Sol' signifies that the terminal is suitable to accept only a solid conductor.

Terminals without these markings are suitable to accept cables of any type.

D 5.3.2 Compression joints

Compression joints and connections are used extensively to make connections for terminating small copper conductors, and for large copper and aluminium conductors.

The use of compression tools generally provides a simple and quick way of making terminations compared with other methods. However, to make a satisfactory termination with a compression tool, care must be taken to select the correct tool so as to:

- match the lug (or other connector) to the conductor; and
- match the compression tool to the lug.

Generally, is it better to choose a lug and compression tool from the same manufacturer and range to ensure compatibility. The matching of cable diameter to lug size is critical and careful inspection of manufacturers' compatibility data is required as the cable overall diameters can vary.

Where the lug and conductor are not compatible, or where the lug and compression tool do not match or are of different manufacture, problems may occur. In the case of a lug and conductor mismatch, attempts to solve the problem by, for example, 'filling' the lug with additional copper strands is not acceptable.

D 5.3.3 Galvanic (electrolytic) corrosion between metals

Where certain dissimilar metals are placed in contact with each other, precautions should be taken to prevent galvanic (also called electrolytic or dissimilar metal) corrosion.

There are two conditions that cause this corrosion and both need to exist for it to take place. Firstly, there have to be two different metals present, and these need to have different atomic properties; this is quantified by the metals' anodic indices. Table D 5.2 shows the anodic indices for metals used in electrical installation engineering. Secondly, there must be an electrical conductive path between the metals. The amount of conduction depends upon water or humidity content as well as chemical presence. To make things a little easier, Table D 5.1 details recommended maximum differences in anodic indices for different environments.

Table D 5.1 is used with anodic indices of metals specified in electrical installation engineering, given in Table D 5.2.

Table D 5.1 Recommended anodic difference for certain environments.

Environment type	Recommended maximum difference in anodic index (V)	Typical example environments
Normal indoor	0.5[1]	All heated non-damp buildings, air conditioned buildings, humidity controlled buildings and structures
Unheated buildings	0.25	Non-humidity controlled warehouses, unheated buildings such as churches and barns, etc.
Outdoor and harsh installations	0.15	Outdoor installations, 'salty' indoor environments, high humidity or moisture

[1] Note that the figure of 0.5 for indoor installations is a design value and there are many installations with metals coupled with anodic differences of 0.7 and 1.0 that show no signs of corrosion after many years of service.

Table D 5.2 Anodic index of common metals.

Metals	Anodic index (V)
Nickel	0.30
Copper, silver solder	0.35
Brass, bronze	0.35
Chromium	0.5
Tin, lead tin solder	0.65
Iron and steels	0.85
Aluminium	0.9
Zinc	1.25

A remedy for making a connection is often plating the metal. For example, as can be seen, copper can be mated indoors with steel or tin (difference of 0.5V and 0.3V respectively). If aluminium needs to be connected indoors to brass or copper, the anodic difference is 0.55; a solution is to 'tin' either the brass or copper which reduces the anodic difference to 0.25.

D 5.3.4 Accessibility of connections

Every connection between conductors, and between a conductor and an item of current-using equipment, is required to be accessible. This is required in order that inspection, testing and maintenance work can be undertaken during the lifetime of the installation. However, there are certain exceptions permitted by Regulation 526–3 where this requirement can be relaxed, such as:

- a compound-filled or encapsulated joint;
- a connection between a cold tail and a heating element (e.g. a ceiling and floor heating system, a pipe trace-heating system);
- a joint made by welding, soldering, brazing or compression tool;
- a joint forming part of the equipment complying with the appropriate product standard.

D 5.3.5 Temperature of connections

Generally, the temperature of connections is not normally an issue where good workmanship and good materials are used. However, there are some circumstances where the temperature attained is of concern. For example, this would be the case of a connection of a conductor insulated with PVC (70°C) to a busbar which is rated for a maximum operating temperature of 90°C, and which is to operate at that temperature.

Without measures to address the problem, the insulation will be impaired (Regulation 526.4). A solution would be to strip back the insulation to a point where the temperature of 70°C is not exceeded. Depending on the temperature gradient, the stripping back of, say, 150 mm of the PVC insulation and replacing it with suitable heat-resisting insulation would solve the problem.

For connections to luminaires, a new section has been added to BS 7671: 2008, Section 559, see D 14.

D 5.3.6 Enclosures of connections of live conductors

Connections and joints in live conductors (and in a PEN conductor) are required to be made within one of the enclosures (or a combination of) identified in Regulation 526.5 as follows:

- a suitable accessory complying with the appropriate product standard;
- an equipment enclosure complying with the appropriate product standard;
- an enclosure partially formed or completed with building material which is non-combustible when tested to BS 476–4.

An enclosure used to contain connections or joints is required to exhibit adequate mechanical protection and be suitable for the external influences likely to be present (Regulation 526.7). Also, the connections are required not to be subjected to any appreciable mechanical strain, i.e. the conductors must be fixed in the vicinity of the enclosure (Regulation 526.6).

Additionally, cores of insulated-only conductors, including those of stripped-back sheathed cables, are required to be contained within the enclosure (Regulation 526.9).

D 5.3.7 Connections – Multi, fine and very fine wire

Regulation group 526.8 is new for BS 7671: 2008. It requires special treatment of connections of multiwire, fine wire and very fine wire conductors and requires that suitable terminals be used or the conductor ends be suitably treated. Regulation 526.8.2 prohibits the soldering of fine wire conductor ends where screw terminals are used.

Soldered conductor ends on fine wire and very fine wire conductors are not permitted (Regulation 526.8.3) where relative movement can be expected between soldered and non-soldered parts of the conductor.

D 5.4   Wiring systems – minimizing spread of fire

Section 527 deals with minimizing the spread of fire and calls for the wiring systems to be installed such that the building structure is not compromised in terms of performance and fire safety.

D 5.4.1 Cable specification

The key regulation in Section 527 is 527 1.3, which states that cables should comply with BS EN 60332–1-2.

> *Tests on electric and optical fibre cables under fire conditions. Test for vertical flame propagation for a single insulated wire or cable. Procedure for 1 kW pre-mixed flame.*

Cables manufactured to this standard (the 'flame test' standard) are considered to be suitable for installation without any special precautions.

It should be noted that all BS EN installation cables comply with this standard except butyl rubber flexes and hence these can only be used in short lengths. Alternatively, cables not complying with BS EN 60332–1-2 can be used in cable management products; these should be to one of the following standards:

● BS EN 61386 *Conduit systems for cable management. General requirements.*
● BS EN 500854: *Cable trunking and ducting systems for electrical installations.*

D 5.4.2 Sealing building penetrations

Regulation group 527.2 requires that penetrations made in building fabric should be made good with a material that maintains the fire rating of the surrounding fabric. Furthermore, internal fire sealing is required for metal and non-metal cable containment systems; the only exception is internal sealing of 25 mm diameter and smaller conduits. This requirement has proven to be most costly for organizations that did not price accordingly, so beware.

Similarly, sealing which has been disturbed by any alteration work is required to be reinstated as soon as practicable.

Under fire conditions, the toxic effects of PVC have long been acknowledged as problematic when considering the effective escape of persons from buildings.

The importance of the selection of PVC on an installation is outside the scope of this book as it will depend upon the view of the local authority. However, with the increasing availability of low smoke, halogen-free (LSHF) cables, the designer can choose to limit PVC or not use it at all.

Bulk PVC cable installations should not be used in designated fire escape zones.

It should be noted that BS 7671: 2008 uses the term 'thermoplastic' instead of PVC and 'thermosetting' generally for LSHF cables.

Examples of cables with low emission of smoke and corrosive gases under fire conditions are the following:

● non-sheathed single-core thermosetting insulated;
● armoured thermosetting insulated;
● elastomer insulated (types C and D only), armoured and non-armoured;
● mineral insulated (except those served PVC overall) with fittings to BS 6081.

Some less common types of cable are given below:

- impregnated paper-insulated aluminium-sheathed (CONSAC) to BS 5593 (often used for PME power supplies);
- PVC insulated cables for switchgear to BS 6231 (used for switchgear);
- impregnated paper-insulated lead-sheathed to BS 6480 (used for HV supplies);
- rubber insulated flexible to BS 6708 (for use in mines and quarries);
- rubber insulated cables and cords to BS 6726 (used for festoon and temporary supplies);
- flexible cables to BS 6977 (used for lift installations).

Cables manufactured to other recognized Standards, provided they meet an equivalent degree of safety.

D 5.5 Proximity to other services

Section 558 of the Regulations includes a few regulations on the subject of proximity to other services and is covered in this book under Section D 4.

D 6 Circuit breakers

D 6.1 General

There is a certain overlap in this chapter with circuit breakers mentioned in Chapter C, which covers the selection of circuit protective devices to co-ordinate with cable sizes.

This section reviews the other selection aspects of circuit breakers as well as their operation, and should be read in conjunction with Chapter C.

There are many types of circuit breaker available, the most common being the thermal magnetic circuit breaker. The three types of circuit breaker in use are BS EN 60898, BS EN 61009 and BS EN 60947. The 60898 devices generally replace the miniature circuit breaker (MCB). 'Miniature circuit breaker' is a deprecated term and these devices should all be called 'circuit breaker'. To avoid confusion these are referred to in parts of this book either as MCBs or occasionally as '60898 devices'.

- BS EN 60898: *Circuit breakers for overcurrent protection for household and similar installations.* These are available in ratings from 6 A to 125 A. They have three different magnetic sensitivities, described in D 6.2 below.
- BS EN 61009: *Residual current circuit breakers with integral overcurrent protection for household and similar uses.* These are combined circuit breakers/residual circuit breakers, known as RCBOs. They are generally available in the same range as 60898 devices, with the same overcurrent sensitivities as 60898 devices and residual sensitivities as RCDs.
- BS EN 60947 circuit breakers replace what were known as MCCBs and air circuit breakers.

Most circuit breakers are available in one-, two-, three- and four-pole variations. The two-, three- and four-pole are all additionally available with neutral overload sensing.

The basic, important requirement for the co-ordination of circuit breakers with cables and load current, as explained in some detail in Chapter C, is given by the formula:

$$I_b \leq I_n \leq I_z$$

D 6.2   Operation and characteristics

An MCB is a thermo-magnetic device, meaning that it has two methods of circuit interruption. A thermal mechanism, usually a bi-metallic strip, provides protection against moderate overcurrent. The heating action of the current causes the bi-metallic strip to curve and break circuit contact. This method is complemented by a solenoid designed to respond to larger currents.

A diagram of an MCB is shown in Figure D 6.1.

It should be apparent that the thermal trip has a slow response time and the solenoid trip has a rapid response time. When combined, these devices provide quite a sophisticated protection characteristic profile. The two characteristics are now described.

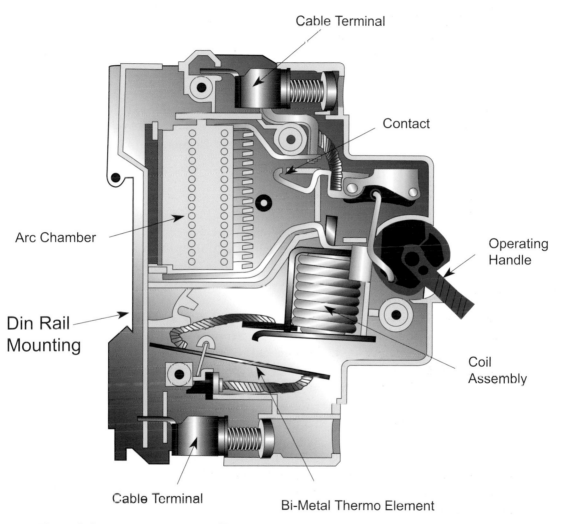

Cable Terminal

Contact

Operating
Handle

Arc Chamber

Din Rail
Mounting

Coil
Assembly

Cable Terminal

Bi-Metal Thermo Element

Figure D 6.1 Internal view of an MCB.

Table D 6.1 BS EN 60898 device thermal characteristics.

Current	Desired result
1.13 I_n	Must not trip within 1 h
1.45 $I_n \leq 63A$	Must trip within 1 h
1.45 x $I_n > 63A$	Must trip within 2 h
2.55 $I_n \leq 32A$	Must trip between 1 and 60 s
2.55 $I_n > 32A$	Must trip between 1 and 120 s

Thermal characteristic

The thermal, bi-metallic characteristic is summarized in Table D 6.1.

A further co-ordination of the requirement is that of Regulation 433.1.1 (iii) which is:

$$I_2 \leq 1.45 \times I_z$$

where I_2 is the current that causes operation of the device. By studying Table D 6.1, it can be seen that this requirement is built into the product standard for BS EN 60898 devices and is effectively the calibration of the bi-metallic strip.

Magnetic characteristic

The maximum rated current available for MCBs is 125 A, and these BS EN 60898 devices are available with different magnetic sensitivities, denoted with a prefix B, C or D accordingly. The different magnetic characteristics of BS EN 60898 circuit breakers are provided in Appendix 3 of BS 7671: 2008, but to illustrate the differences in the magnetic characteristics, Figure D 6.2 shows a comparison of B, C and D types for devices of the same basic rating.

A 32 A circuit breaker with type C sensitivity is denoted C32, and it is a requirement of the equipment standard to apply this marking to the device.

The stated B, C or D sensitivities each have a minimum current that causes operation, and this is conventionally taken to be operation within 0.1 second; this is conventionally termed instantaneous operation or instantaneous tripping. This minimum time convention is due to the mechanics of the circuit breaker, which will always require a certain minimum time, regardless of current for the trip mechanism to open.

This instantaneous tripping time of 0.1 second can be a problem where two devices are required to discriminate, and this is also explained in Chapter C.

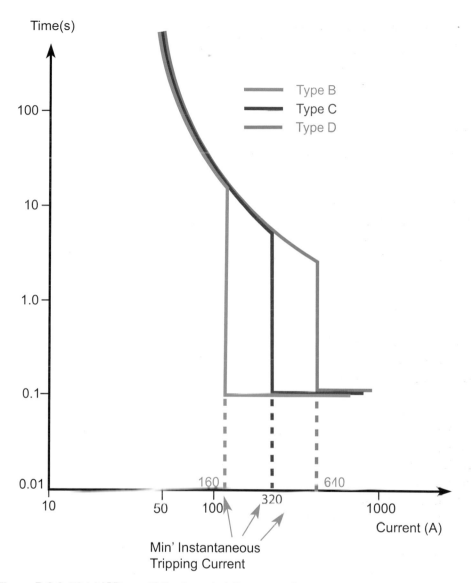

Figure D 6.2 32 A MCB sensitivity characteristics comparison.

Figure D 6.2 shows that in order to achieve instantaneous tripping or tripping at 0.1 s, a 32 A type B breaker requires 160 A, a type C breaker 320 A and a type D breaker 640 A.

Table D 6.2 Circuit breaker (BS EN 60898) selection for inrush current applications.

Type	Manufactured magnetic trip setting (× I_n)	Typical applications
B	3 to 5	General domestic and resistive loads
C	5 to 10	Small motors (a few kW), small transformers fluorescent lighting and most inductive loads
D	10 to 20	DOL motors, large star delta motors, low-pressure sodium discharge lighting, larger transformers, welding machine supplies

Below these threshold currents the thermal mechanism is dominant, and has the same characteristic for all three devices. The magnetic characteristics determine the sensitivity type. Equipment connected or likely to be connected to the circuit must be assessed in terms of likely peak or inrush current (533.2.1).

Inrush current is the current that a load draws when the supply is switched on. Values can range from being insignificant (a few times the normal current), 5 to 10 times normal current for iron core transformers (e.g. conventional ballast-fluorescent luminaires) and up to 20 times normal current for much modern electronic equipment, including the power supplies found in user equipment. While short-lived (often the peak current is a few milliseconds), this can cause circuit breakers to trip, but assessing the likelihood of a circuit breaker tripping is complicated. Table D 6.2 recommends circuit breaker types for typical inrush current applications.

Separately to inrush current, load peak current also needs to be considered. Peak current in respect of circuit breaker selection is a term used to describe a peak within the normal operation of a cyclic or time varying load. Part C 3.1 of this book has more information and an example of a cyclic load calculation. If you have loads with significant cyclic peaks you need to confirm that the circuit breaker will not trip. This can be confirmed by studying the circuit breaker characteristic curve, but confirmation with the manufacturer may be necessary.

D 6.3   Ambient temperature de-rating

Both BS EN 60898 and BS EN 60947 provide details of de-rating factors for using circuit breakers in ambient temperatures greater that 30°C.

In installations where you encounter such temperatures it is best to consult the standards or manufacturers' data, but the following give an indication of correction factors for 60898 devices up to 100A:

- 40°C de-rating factor of 0.85
- 50°C de-rating factor of 0.75
- 60°C de-rating factor of 0.65.

Some manufacturers also provide de-rating factors for groups of circuit breakers mounted next to each other where they are all fully loaded. A typical de-rating in such examples would be 0.85, but in practice it is very rare to see such loadings, particularly on more than one circuit.

D 7   Residual current devices

D 7.1   BS 7671 applications

BS 7671: 2008 requires the fitting of RCDs as additional protection in more applications compared with the previous editions. RCDs, with a rated operating current of 30 mA, are now required in the following situations:

- All socket-outlet circuits accessible to ordinary or non-instructed persons (411.3.3).
- All circuits within a bathroom (701.411.3.3).
- Concealed cables either less than 50 mm deep, in installations not intended to be under the supervision of a skilled or instructed person, not mechanically protected or protected by earthed metal and not run in the 'safe zones'.
- Concealed cables within metal partitions, in installations not intended to be under the supervision of a skilled or instructed person, not mechanically protected or protected by earthed metal and accessible to ordinary or non-instructed persons.
- Generally where disconnection times cannot be achieved with an overcurrent device.
- Most TT installations, usually required to achieve a disconnection time of 0.2 s.
- Other special locations including caravan parks.

One of the most significant changes in the 17th Edition is the requirement for RCDs to be applied to all socket-outlet circuits accessible to ordinary persons. It is quite important to get the principle of Regulation 411.3.3 clear.

411.3.3 Additional Protection

In a.c. systems, additional protection by means of an RCD in accordance with Regulation 415.1 shall be provided for:

(i) socket-outlets with a rated current not exceeding 20 A that are for use by ordinary persons and are intended for general use, and

(ii) mobile equipment with a current rating not exceeding 32 A for use outdoors.

An exception is permitted for:

(a) socket outlets for use under the supervision of skilled or instructed persons, e.g. in some commercial or industrial locations, or

(b) a specific labelled or otherwise suitably identified socket-outlet provided for connection of a particular item of equipment.

Note 1: See also Regulations 314.1(iv) and 531.2.4 concerning the avoidance of unwanted tripping.

Note 2: The requirements of Regulation 411.3.3 do not apply to FELV systems according to Regulation 411.7 or reduced low voltage systems according to Regulation 411.8.

Let's break down the requirement and also describe ordinary, skilled and instructed persons.

An ordinary person is a person who is neither skilled nor instructed. A skilled person is one with technical knowledge or sufficient experience to enable him or her to avoid dangers which electricity may create. An instructed person is one adequately advised or supervised by a skilled person to enable him/her to avoid dangers which electricity may create.

Applying this to Regulation 433.1.1 (and elements of 522.6.7 and 522.6.8):

Domestic installations always involve ordinary persons, and RCDs are required to all socket-outlet circuits. Exceptions can be made for socket outlets for specific purposes, a common example of which would be a domestic freezer circuit. To comply with the spirit some of these socket outlets may need labelling depending upon their location.

Commercial installations where individuals are employed come under the Electricity at Work Regulations 1989 (EWR). This is also true of a single employee workplace whatever the nature of the business. EWR requires a safe system of working, and this means that someone in every installation will need to control activities; the operational requirements of installations are made in BS EN 50110: 2004–1: *Operation of Electrical Installations*. When this requirement is combined with the requirement under EWR for the workplace equipment to be safe and maintained, there is no requirement for employees in the workplace to have socket outlets with additional RCD protection. The designer must assume this to be the case.

Falling outside of these two cases are the commercial installations with socket outlets that members of the public can use. These would not be socket outlets in the lobby of a commercial installation, where it is reasonable to assume they will not be used by visitors, but would include, say, socket outlets in internet cafés for plugging in laptops. The designer has options and will need to liaise with the client over supervising or providing instruction to such users or fitting RCDs to socket outlets.

D 7.2 Operation and BS 7671 requirements

Operation

Residual current devices monitor the instantaneous current flowing along the line and neutral conductors in a circuit by means of a sensitive coil. The coil or 'core' measures the instantaneous resulting sum of these currents. When an earth fault or earth leakage is present on a circuit, a small 'imbalance' is set up in the detection coil and, at a certain level, causes a trip of the circuit breaker mechanism. A diagrammatic representation of this is shown in Figure D 7.1.

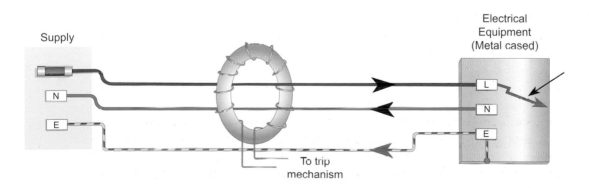

Figure D 7.1 Representation of an RCD.

Requirements

The requirements for RCDs are set out in Regulation group 531.2 and are summarized in Table D 7.1.

Table D 7.1 Requirements for RCDs.

Reg No	Requirement
531.2.1	Capable of disconnecting all the line conductors of the circuit at the same time
531.2.2	Magnetic circuit to enclose all live conductors and protective conductor to be outside the magnetic circuit
531.2.3	Residual operating current to comply with the requirements of Section 411 (automatic disconnection)
531.2.4	Selected so any expected protective conductor current will be unlikely to cause unnecessary tripping of the device
531.2.5	Not be considered sufficient for fault protection even if the rated residual operating current does not exceed 30 mA
531.2.6	Where powered from an independent auxiliary source must be fail safe
531.2.7	Not be impaired by magnetic fields caused by other equipment
531.2.8	Where used for fault protection separately from an overcurrent device, must be capable of withstanding thermal and mechanical stresses
531.2.9	Where discrimination is necessary to prevent danger, device characteristics are required so discrimination is achieved
531.2.10	For ordinary persons it shall not be possible to modify or adjust the setting of the device without a tool or key

Most of the above requirements will be achieved automatically as they are incorporated within the relevant product standards. This is not so for the requirements for 'unwanted tripping' or discrimination, and these are now discussed.

D 7.3   Unwanted RCD tripping and discrimination

Unwanted tripping

In order to avoid unwanted tripping, the most important factor of influence is the circuit arrangement. Most modern individual pieces of equipment on a circuit create some amount of earth leakage current and these currents add in the circuit's protective earth conductor. There is, therefore, a circuit limit to the quantity of connected equipment in terms of the total circuit earth leakage. In order to reduce the risk of unwanted tripping, the circuit should be designed so that the circuit runs with a standing earth leakage of less than 25 % of the rated residual operating current of the RCD.

Two common sources of equipment earth leakage are as follows:

1   Earth leakage caused by power supplies, particularly switched mode power supplies. These are used in a considerable amount of modern equipment including computers and televisions, photocopiers etc. A rule of thumb figure of 1 mA leakage per device is recommended for assessment.

2   Heating elements found in washing machines, dishwashers and cookers. A rule of thumb is 3 mA for appliances up to 3 kW and 6 mA for appliances up to 6 kW.

For speculative designs and socket-outlet circuits the designer needs to take a view on likely equipment the user will operate. Generally to achieve compliance with Section 543.7 (high protective conductor currents), a maximum of 10 double socket outlets per circuit is recommended.

Example: a 30 mA RCD circuit with eight PCs, two televisions and a 3 kW dishwasher (a rather strange mix, granted, but this could be a single RCD in a consumer unit). Based upon the above rules of thumb, the earth leakage would be in the order of 13 mA. Under most circumstances, the RCD would not trip even though this exceeds a good design value of 25%. However, if someone now used a domestic kettle in the circuit, the earth leakage may trip the RCD.

This example should serve to illustrate the principle of RCD selection and its effect on circuit design.

There are many other factors that may either contribute to or cause unwanted tripping, including inrush currents (see Section D 6.2 for explanation), mains factors such as noise, loose connections, poor insulation of cables of connected equipment and radio frequency interference.

There are other causes of unwanted tripping. A relatively rare problem, but one that is worth mentioning, is unwanted tripping that is caused by microwave emissions from mobile cell phones. This can be an issue and exposes a loophole in standardization. It is a valid reason in itself not to install RCDs on safety-of-life critical circuits, along with suitable comments made on the completion certificate.

RCD discrimination

In order to achieve discrimination between two series-connected residual current devices where required to comply with Regulation 531.2.9, it will be necessary to use a time-delayed residual current device at the upstream position. This will permit the downstream residual current device to respond to an earth fault on its downstream side and avoid operation of the upstream device.

D 7.4 d.c. issues for RCDs

The selection of RCDs in respect of load d.c. components is an issue that is often overlooked by designers. RCDs are classified according to their response to d.c. signals as follows:

- **Type AC** This class of device generally only detect sinusoidal alternating residual currents. They may not detect non-sinusoidal, non-alternating residual components. These non-sinusoidal currents are present in many items of equipment, e.g. virtually all equipment with a switched mode power supply will have a d.c. component, as do battery chargers and X-ray machines etc.
- **Type A** This class of device will detect residual current of both a.c. and pulsating d.c. and are known as d.c. sensitive RCDs. They cannot be used on steady d.c. loads.
- **Type B** This type will detect a.c., pulsating d.c. and steady d.c. residual currents.

RCDs are required to be marked with their type and Table D 7.2 shows the markings and provides selection criteria.

D 7.5 TT installations and RCDs

In a TT system the earth fault loop impedance is not usually sufficiently low to facilitate the operation of an overcurrent protective device within the required disconnection time. This is even more the case in BS 7671: 2008 as the disconnection time for TT installations is 0.2 s compared with the 0.4 s required in the previous edition.

Where an RCD is used for fault protection, the product of the rated residual current and the sum of the resistances of the installation earth electrode and protective conductors connecting it to exposed-conductive-parts (R_A) is required to be not greater than 50 V (Regulation 411.5.3). However, it is also recommended that an earth electrode resistance, where possible, should not exceed 200 Ω (411.5.3 note 2). This is good general advice but is a simplification, as a deep earth electrode, say 4 m deep with resistance of 400 Ω, will be more stable than a 1.2 m electrode with resistance of 50 Ω. For rod electrodes it is vertical depth that should be encouraged.

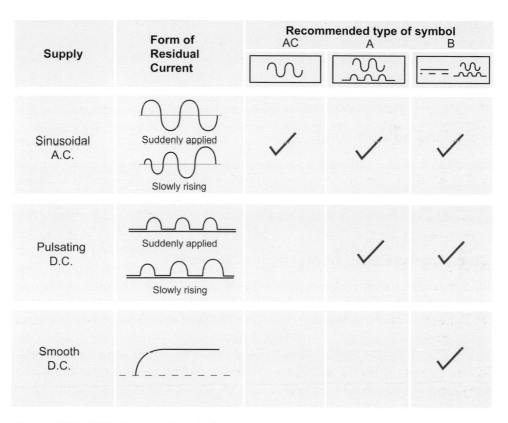

Supply	Form of Residual Current	Recommended type of symbol		
		AC	A	B
Sinusoidal A.C.	Suddenly applied / Slowly rising	✓	✓	✓
Pulsating D.C.	Suddenly applied / Slowly rising		✓	✓
Smooth D.C.				✓

Figure D 7.2 RCD d.c. selection criteria.

In a TT installation, where a single RCD protects the whole installation, it is essential that the device is placed at the origin of the installation (i.e. adjacent to the electricity meter). The connections and any switchgear between the RCD and the electricity supplier's meter is required to meet the requirements for protection by the use of Class II equipment or equivalent insulation (531.4.1).

D 8   Other equipment

There are a number of regulations grouped under Section 530 listed as protection, isolation, switching, control and monitoring. This section lists the most relevant points; most others do not need commenting on.

D 8.1   Isolation and switching

The ECA has always found it disappointing that the subject of isolation and switching was so convoluted to the point that there was a separate IEE guidance note specifically on the subject. A new Table 53 within Chapter 53 answers many questions posed by designers or installers.

D 8.2   Consumer units for domestic installations

BS EN 60898 circuit breakers installed in domestic consumer units would not generally comply with Regulation 432.1 where a breaking capacity of 16 kA is quoted. However, a consumer unit that complies with the conditional short-circuit test as described in Annex ZA of BS EN 60439–3 will meet the requirements of the Regulations as compliance with 434.5 is achieved. This means that where a service cut-out fuse to BS 1361 with a rating of maximum 100 A is fitted the switchgear will be safe. This is stated here merely from a specification point of view.

D 8.3   Overvoltage, undervoltage and electromagnetic disturbances

Chapter 44 affects the selection of equipment. The chapter has been extensively modified, for the 17th Edition gives requirements for protection against voltage disturbances and electromagnetic disturbances. It considers these disturbances for both internal and externally created disturbances, and is divided into four sections; two are 'blanks' for BS 7671: 2008 as follows:

- Section 441: Reserved for future use.
- Section 442: Protection of LV installations against HV faults.
- Section 443: Protection against overvoltages of atmospheric origin or due to switching.
- Section 444: Measures against electromagnetic influences. At the time of publication this section was being developed by the Wiring Regulations committee and will be published in a future amendment.

D 8.3.1 HV – LV faults

The new Section 442 sets regulations for installations for supplies taken at high voltage or for where a substation is located on the property of the installation.

Section 442 specifics a maximum 'stress voltage', which is a fault voltage appearing on the LV earthing system with respect to the local earth potential.

However, installations that receive their LV supply from a utility provider do not need to consider this section. Thus, Section 442 can be ignored unless you are installing a private substation. In these cases reference to IEC 61936–1 *Power Installations Exceeding 1 kV a.c.* will provide methods of complying with the HV/LV co-ordination aspects.

D 8.3.2 Atmospheric and switching overvoltages

Like many parts of the Regulations, there is often confusion when reading a section, only to discover later that most of the text is not applicable. Section 443 is a classic case of this and is explained as follows:

● No action will be required for UK installations as installations will be in areas where thunderstorms occur less than 25 days per year.

It is noted that equipment will need to have voltage impulse withstand levels as specified in Table 44.3, but all equipment to BS EN standards will comply with this. For installations overseas in areas of high lightning strike locations, surge protection is required (alternatively a risk assessment may be carried out).

D 8.3.3 Undervoltage

Although not new, the regulations under Section 445 required an undervoltage detection relay with appropriate resetting for installations where a sudden re-energized state could lead to danger. Common examples of these installations are machinery workshops and similar plants, and there are various standards to consider, the subject of which is outside the scope of this book.

D 8.4 Surge protective devices

Although not required in the UK by Section 443 (see Section D 8.3.2), surge protective devices (SPDs) may optionally be specified. If you are the designer you may feel that they add a certain security to an installation. SPD selection and installation comes down to a few simple rules, summarized as follows:

● Choose SPDs complying with IEC 61643 series.
● In TN-C-S installations connect device between line conductors and earth.
● In TN-S and TT installations connect device between line conductors and both neutral and earth.

Figures D 8.1, D 8.2 and D 8.3 show typical SPD connections in mains positions of various supply earthing arrangements.

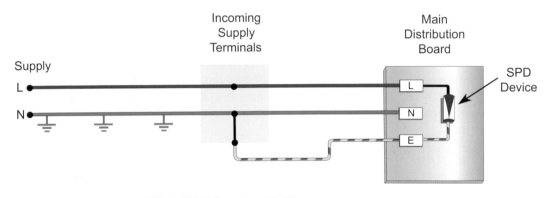

Figure D 8.1 Connection of SPD in TN-C-S system (PME).

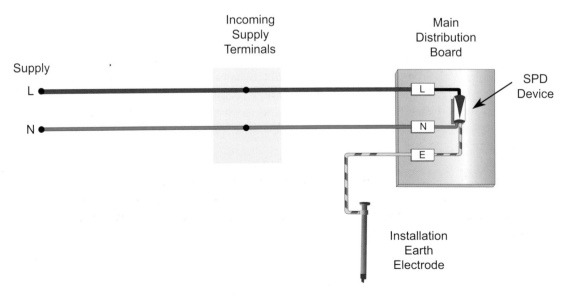

Figure D 8.2 Connection of SPD in TT system.

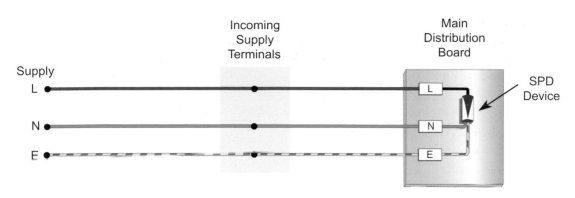

Figure D 8.3 Connection of SPD in TN-S system.

The length of conductors connecting the SPDs is a factor that should be controlled due to the magnitude of both current and voltage when the devices operate, and it is recommended that connecting tails are limited to a length of 0.5 m.

D 8.5 Insulation monitoring devices (IMDs)

Most of BS 7671: 2008's rules (538.1) for insulation monitoring devices (IMDs) cover their use in IT systems (e.g. for first fault conditions). This is a rare and specialist area and is beyond the scope of this book. Some limited information can be found in BS EN 61557–8.

D 8.6 Residual current monitors (RCMs)

Still a relatively rare device, the residual current monitor (RCM) is similar to RCDs (see D 7.2), but instead of tripping a circuit, monitoring is provided. The monitoring can be visual or, at a pre-determined value of residual current, it can create an audible signal or a signal for remote monitoring. RCMs are useful devices in main distribution panels, and in this configuration monitor several circuits at once.

In a larger installation a three-phase RCM device positioned in the main distribution board will be extremely useful. These devices can be widely used and Figure D 8.4 indicates examples of where RCMs are used in mains, riser and final distribution board positions.

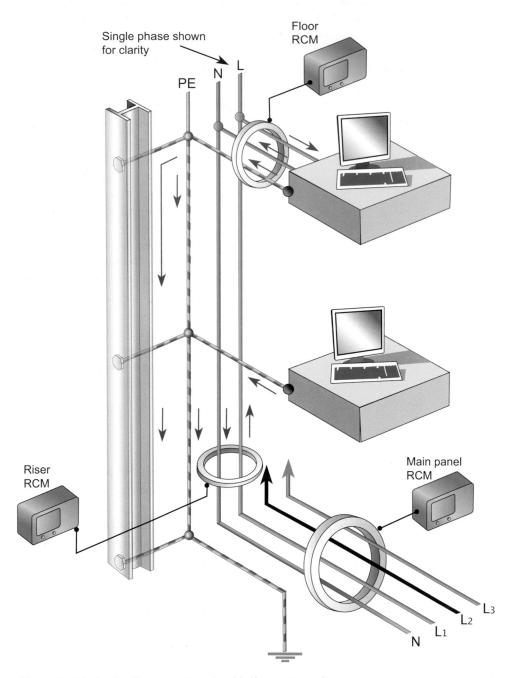

Figure D 8.4 Application examples of residual current monitors.

The following pattern has been experienced by the author in several large installations (load of a few MW) and has been provided here for information for those wishing to use the devices for the first time:

● When setting the RCM, a general standing residual current will be established; this may involve considerable fault finding for faults previously undetected.
● At certain times the residual current will probably suddenly go high; this will often be due to equipment faults, particularly neutral earth faults which remain undetected by overcurrent devices.
● For tracing faults, isolation of sub-boards is the easiest method, but is subject to isolation of large parts of the installation. Where this is not possible the use of clamp-on earth leakage current probe testers will be necessary.
● For tracing faults, user equipment should be subjected to its daily or weekly cyclic run mode.

To serve as an example: a 2 MW installation may have 4 A of 'normal' residual current, fluctuating most days between 3 and 5 A. Suddenly the earth leakage current rises to about 15 A. This may be attributed to a neutral earth fault, present on a lighting circuit or similar.

For smaller installations, some limited information on the devices can be found in BS EN 62020: *Electrical accessories – Residual current monitors for household and similar uses (RCMs)*.

D 9 Generating sets

Section 551 was introduced in the 2001 edition of BS 7671 and has only had limited modification for the 17th Edition.

It is key to recognize the definition of 'generating set' within the scope of Section 551, as it applies to more than just 'diesel' generators. 'Generating set' applies to any power source other than the incoming or main supply, including:

● combustion engines e.g. diesel sets;
● electromotive generators e.g. wind turbines, water mill turbines;
● photovoltaic cells;
● electrochemical accumulators e.g. battery back-up cells.

The key regulations of this section are as in Table D 9.1.

Table D 9.1 Requirements for 'generating sets'.

Voltage/frequency shall not cause damage or danger	551.2.3
RCDs in circuit remain effective	551.4.2
Fault protection (disconnection times) of system must work with various combinations of supply	551.4
Standby supplies to have separate earth independent from mains	551.4.2
Non-permanent sets to have 30 mA RCD	551.4.4.2

The most common non-compliance item here is the requirement to provide an independent earth. However, it is very often easy to comply with this regulation. For most installations the use of structural steel and similar extraneous-conductive-parts can be utilized to form a suitable earth system. Although some checking with Chapter 54 of BS 7671 is required (see Chapter E) often the existing main bonding elements and conductors can be utilized and are more than adequately sized for the purpose.

D 10 Rotating machines

Regulation group 552 addresses rotating machines and makes a number of requirements that are unique to motor circuits, and are in addition to the general requirements, which also have to be met.

Motors of 370 W or more are required to have a device for protection against overload (552.1.2) except where they are part of equipment complying with an appropriate British Standard. Stand-alone motors will normally have a motor starter incorporating an overcurrent device.

Motors with frequent starting and stopping will need a careful rating selection due to the cumulative effect of the high starting current, and manufacturers' data should be considered here. As well as the motor itself, such frequent starting will have an effect on the supply cables and other electrical equipment, which should be de-rated. There is no guidance for such de-rating as the permutations are quite varied, and will have to be considered individually – see the example in Section C 3.1.

Motor control is addressed by Regulation 537.5.4. Where the restarting of a motor after a mains failure could cause danger, a means is required (usually a relay) to prevent the automatic restarting of the motor on re-energization (537.5.4.1).

Regulation 537.5.4.3, whilst not new, is not always complied with and states that where the direction of rotation is a factor relating to safety, measures to prevent the reversal shall be taken. This effectively means supply phase rotation sensing at or near the motor with associated automatic isolation should phase reversal occur.

D 11 Plugs and socket outlets

Regulation group 553 sets out the requirements for plugs, socket outlets and cable couplers in terms of applicable product standards, other particular requirements relating to pin configuration and the need for every socket outlet for household use to be shuttered.

Except for the connection of clocks, electric shavers and other circuits having special characteristics (see Regulation 553.1.5), the only acceptable types of socket outlet for low voltage applications are:

- 13 A plugs and socket outlets to British Standard BS 1363 for a.c. only;
- 2, 5, 15, 30 A fused or non-fused plugs, and socket outlets to BS 546;
- 5, 15, 30 A fused or non-fused plugs, and socket outlets to BS 196;
- 16, 32, 63, 125 A plugs and socket outlets (industrial type) to BS EN 60309–2.

There are many applications where you may wish, or indeed need, to specify or use a plug and socket not listed above. These will generally be 'no-load' make-or-break devices, and are fine for applications where skilled or instructed persons operate.

The mounting height of a socket outlet above floor level, or above a work surface, should minimize the risk of accessory or cord damage (553.1.6). Although not specified as a distance in the Regulations, a height of approximately 150 mm above the floor or work surface is recommended as a minimum.

For installations in domestic premises, Part M of the Building Regulations stipulates mounting heights for electrical equipment including socket outlets (see also BS 8300: *Design of buildings and their approaches to meet the needs of disabled people. Code of practice*).

There has been some confusion with the interpretation of these requirements since their publication. Generally, for both dwellings and 'visitor' areas in commercial installations, electrical accessories should be mounted at a height of between 450 mm and 1200 mm above finished floor level. The basis of these criteria is to avoid problems for disabled persons that may visit a building. It is not a design brief for a dedicated installation for disabled persons; these are quite different.

Finally on socket outlets, Regulation 553.1.7 requires that a sufficient number of socket outlets be provided so that long flexible cords are avoided. This should be pointed out to clients as part of your design briefing.

Cable couplers
Excepting for SELV and Class II circuits, cable couplers are required to be non-reversible and have provision for a protective conductor. Acceptable cable couplers include those complying with the following British Standards:

- BS 196;
- BS EN 60309–2;
- BS 4491;
- BS 6991.

It should go without saying that the connector (female part) of the coupler is fitted to the end of the cable remote from the supply.

D 12 Electrode water heaters and electrode boilers

Regulation group 554.1 sets out the requirements for electrode water heaters and boilers, which are summarized as follows:

- Connected only to an a.c. supply;
- Supply controlled by a linked circuit breaker;
- Overcurrent protective device fitted in each conductor feeding an electrode;
- Shell of electrode to be earthed;
- Where the heater is three-phase, the shell of the heater is required to be connected to the neutral conductor;
- Heaters are required to incorporate (or be provided with) an automatic device to prevent a dangerous rise in temperature.

Generally, all heaters with immersed heating elements are required to have an automatic device to prevent a dangerous rise in temperature. Most of these requirements are built in to the BS and BS EN product standards.

D 13   Heating conductors

With an increase in the specification of underfloor heating, the Regulations remain surprisingly sparse in this area.

The best advice in selecting either underfloor heating cables or external buried heating cables is to select products with recognized product standards and to follow the manufacturers' installation guides. There are no regulations in the group 554–06 which specify anything that would not be included in the product Standards.

D 14   Lighting and luminaires

A completely new section for the 17th Edition is Section 559 'Luminaires and Lighting Installations'.

The new section deals with interior and exterior lighting installations and also applies to highway power supplies and street furniture.

The following items are excluded from Section 559:

- distributors' equipment as defined in the ESQCR 2002[1];
- HV signs supplied at LV, for example neon tubes;
- signs and luminous discharge tube installations operating from a no-load rated output voltage exceeding 1 kV but not exceeding 10 kV.

[1] The Electricity Safety, Quality and Continuity Regulations 2002.

Table D 14.1 summarizes the requirements. Although this table looks fairly long, it covers all the regulations in Section 559 and, consistent with the general spirit of this book, only regulations that are not obvious are included and subsequently explained.

For high voltage signs, see BS 559 *Specification for design, construction and installation of signs* and the BS EN 50107 series *Signs and luminous discharge tube installations operating from a no-load rated output voltage exceeding 1 kV but not exceeding 10 kV.*

Table D 14.1 Lighting and luminaire requirements.

Track lighting shall comply with BS EN 60570	559.4
Heat of lamps on surrounding material shall be considered	559.5.1
Max. lighting circuit size of 16 amp (ES and BC lampholders)	559.6.1.6
Outer screw of ES lampholder must connect to neutral (except for E14 & E27 to BS EN 60238)	559.6.1.8
Through wiring only permitted if luminaire designed for it	559.6.2.1
Temperature rating if marked on luminaire gives heat withstand of cable; see Figure D 14.1	559.6.2.2
3-phase lighting circuits with common neutral shall have linked circuit breaker	559.6.2.3
Stroboscopic effects to be considered in areas with rotating machines (a stroboscopic effect can give the perception that moving objects are stationary when in fact they are moving)	559.9
Earth-free equipotential bonding not to be used	559.10.2
SELV transformers to be to BS EN 61558–2–6[1]	559.11.3.1
ELV electronic convertor to be to BS EN 61347–2–2[2]	559.11.3.2
SELV bare live conductor systems shall comply with BS EN 60598–2–23 or shall monitor and disconnect within 0.3 s, fail-safe	559.11.4.1 and 559.11.4.2
ELV flexes minimum 1.0 mm^2 max. length 3 m	559.11.5.2
ELV flexes minimum 4.0 mm^2 for luminaires suspended by the flex	559.11.5.2
Highway street furniture access door to be accessible only with key or tool or be 2.5 m above ground	559.10.3.1
Highway street furniture lamps to be 2.8 m high or behind barrier	559.10.3.1
Highway street furniture adjacent metal does not require bonding	559.10.3.1
Highway street furniture may use fuse carrier as isolator	559.10.6.1
Highway street furniture disconnection time is 5 seconds	559.10.3.3

[1] and [2] Requirement only applies where SELV is being used as a protective measure. Often installers use SELV equipment but install it to 230 volt standards and this requirement will not apply (e.g. most ELV downlighter installations).

As well as 'throwing the book' at you with quite a few new regulations to study and digest, there is some helpful information in Section 559, mainly the markings found on some luminaires; this information originates in BS EN 60598–1: 2004 *Luminaires* and is given in Figure D 14.1.

SYMBOL	MEANING
□ (square in square)	Class II
$t_a \ldots °C$	Rated maximum ambient temperature
COOL BEAM (crossed out)	Warning against the use of cool-beam lamps
(– – – m|	Minimum distance from lighted objects (metres)
▽F	Luminaire suitable for direct mounting on normally flammable surfaces
▽F (crossed out)	Luminaire suitable for direct mounting on non-combustible surfaces only
▽F (with line above)	Luminaire suitable for direct mounting in/on normally flammable surfaces when thermally insulating material may cover the luminaire
$t \ldots °C$ (cable symbol)	Use of heat-resistant supply cables, interconnecting cables or external wiring
(bowl mirror lamp symbol)	Luminaire designed for use with bowl mirror lamps
⊤	Rough service luminaire
△E	Luminaire for use with high pressure sodium lamps that require an external ignitor (to the lamp)
△I	Luminaire for use with high pressure sodium lamps having an external starting device
(Rectangular) or (Round)	Replace any cracked protective shield
(self-shielded lamp symbol)	Luminaire designed for use with self-shielded tungsten halogen lamps only

Figure D 14.1 Luminaire markings from BS EN 60598.

D 15 **Safety services**

D 15.1 Introduction

Chapter 56 of BS 7671: 2008 entitled 'Safety Services' covers some basic selection and erection aspects of supplies intended for safety services. A safety service is defined as an electrical system provided to protect or warn persons in the event of a hazard, or essential to their evacuation.

Chapter 56 sets some fundamental requirements only, and there will be considerable overlap with other safety service related standards in both design and installation issues. Most of the chapter provides basic rules about standby supplies where used for safety services.

The requirements of the chapter again have been broken down into tables, one for equipment selection and erection and the other for circuits. Obvious regulations have not been included.

D 15.2 Classification of break times

For an automatic safety service, the classification of changeover durations is summarized in Table D 15.1.

Table D 15.1 Classification of changeover durations for safety supplies.

Classification	Duration 'D' of changeover (s)	Break description
No-break	Zero	An automatic supply which can ensure a continuous supply within specified conditions
Very short break	$D \leq 0.15$	Changeover within stated duration
Short break	$0.15 \leq D \leq 0.5$	Changeover within stated duration
Lighting break	$0.5 \leq D \leq 5$	Changeover within stated duration
Medium break	$5 \leq D \leq 15$	Changeover within stated duration
Long break	$15 \leq D$	Changeover within stated duration

Note: In order to maintain the specified operation, essential items of equipment for safety services are required to be compatible with the changeover duration as indicated above.

D 15.3 Safety sources

There are a number of requirements peculiar to safety sources, these being summarized in Table D 15.2.

Table D 15.2 Summary of requirements for sources for safety services.

Type	Regulation	Requirement
Sources	560.6.2	Source to be accessible only to skilled persons or instructed persons
	560.6.4	Independent supply feeders by distribution network operator (DNO) not allowed unless DNO confirms feeders unlikely to suffer simultaneous failure
	560.6.6	Source may be used for purposes other than safety services if service is not impaired
		A fault occurring in other than a circuit for purposes of safety services is required not to interrupt any circuit for safety services
		To achieve above, automatic load shedding will often be required
Non-parallel sources	560.6.7	Precautions required to prevent parallel operation
Parallel sources	560.6.8	The parallel operation of independent sources usually requires the authorization of the distributor
	560.6.8.1	Protection against fault current and against electric shock is required to be ensured where the installation is supplied separately by either of the two sources or by both in parallel
Central power supply sources	560.6.9	The batteries are required to be vented or valve-regulated type. Minimum design life of the batteries to be 10 years
Low power supply sources	560.6.10	The batteries may be of gastight or valve-regulated maintenance-free type. Minimum design life of the batteries is required to be 5 years
UPS	560.6.11	UPS must be able to operate protective devices
Generator supply sources	560.6.12	Must comply with BS 7698–12: 1998

D 15.4   Circuits for safety services

Circuits for safety services are subject to additional requirements, which are summarized in Table D 15.3.

Table D 15.3 Additional requirements for circuits for safety services.

Regulation	Additional requirement/comment
560.7.1	To be independent of other circuits
	If routed in fire risk areas, fire rating required
560.7.2	Routing through fire risk areas should be avoided where practicable, if not possible fire rating required
560.7.3	Overload protection may be omitted where the loss of supply may cause a greater hazard (indication required e.g. visual or audible)
560.7.4	Overcurrent protective devices to be co-ordinated so as to achieve discrimination
560.7.5	Switchgear and control gear accessible only to skilled persons or instructed persons
560.7.8	Circuits for safety services not to be installed in lift shafts or other flue-like openings
560.7.11	All current-using equipment to be listed, together with rated currents and starting currents and time. This information may be included in the circuit diagrams
560.8.1	Fire rated cables to be to BS EN 50362, BS EN 50200, BS EN 60332–1-2
560.8.2	Where applicable control bus must comply with above (all circuit requirements)

D 16   Ingress protection (IP), external influences

D 16.1   General

Regulation 512.2 requires that equipment is suitable for the external influence that it will be subjected to when installed. A large 'matrix' type table of external influences with letter designations is included in Appendix 5 of BS 7671: 2008, running to 13 pages. This can be confusing to use and is largely used by technical committees.

The influence of objects and water is the main criterion that needs to be considered; other criteria, like ambient temperature, are best found from studying manufacturers' data.

The European Standard BS EN 60529: *Degrees of protection provided by enclosures (IP Code)* is a 'standard for standards' document and provides information about the degrees of protection that can be expected of equipment when allocated with a particular IP Code.

'Standard for standards' means that the document is intended for use by equipment manufacturers and equipment standards committees.

The IP code provides an indication of the protection the equipment will afford in terms of access to hazardous parts, such as live or moving parts, ingress of foreign objects such as tools, dirt, and liquids such as water. Exact confirmation on particular performance should always be made with the manufacturer.

The arrangement of the IP code is made up of four characters, some of which are optional. The arrangement of the code is as follows:

IP- International
   Protection

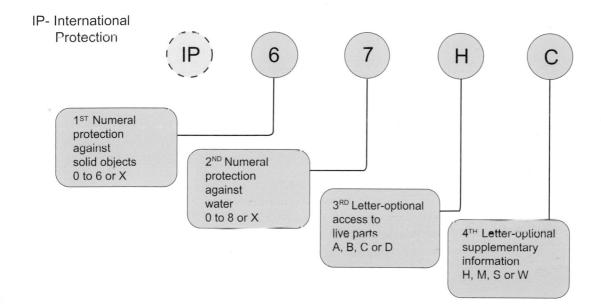

Now the code is established, the protection levels can be understood and these are often only specified using the first two characters, as follows:

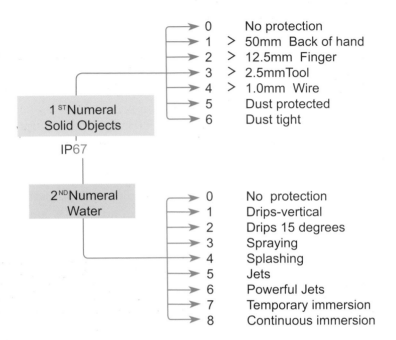

Sometimes, albeit rarely, the optional characters three and/or four may be used as follows:

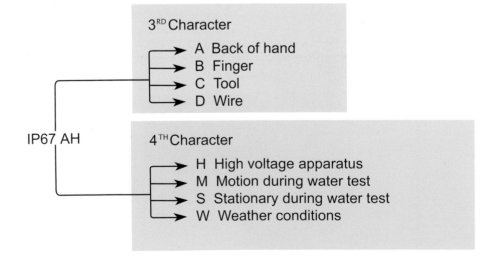

D 16.2   Equipment applications and examples

Sometimes, particularly in specifications, one of the first or second characters is not specified and an 'X' is used, e.g. 'IP2X'.

Perhaps a little more confusing is where the two designations are used for water. Generally, the higher the numerals of the IP code, the higher the protection level. This is not true where jets of water are concerned. Equipment to IPX8, for example, is totally immersible in water but may not necessarily be suitable for jets (IPX5) or high-pressure jets (IPX6). Equipment that is suitable for both environments will thus require a dual marking of IPX6/IPX8 meaning protected against powerful jets and able to be immersed.

Table D 16.1 applies to water jets and immersion.

Table D 16.1 Water jets and immersion.

Enclosure passes tests for:			
Water jets second characteristic numeral	Temporary or continuous immersion second characteristic	Designation and marking	Range of application
5	7	IPX5/IPX7	Versatile
6	7	IPX6/IPX7	Versatile
5	8	IPX5/IPX8	Versatile
6	8	IPX6/IPX8	Versatile
—	7	IPX7	Restricted
—	8	IPX8	Restricted

It can be seen that equipment with the single designation IPX7 or IPX8 has restricted applications or use.

Finally, the following symbols are not listed in BS EN 60529 but may be encountered; an equivalent of the IP code is given below and this will not always be reproduced on the equipment.

⬤	Drip Proof	(IPX2)
⬛	Rain Proof	(IPX3)
▲	Splash Proof	(IPX4)
▲ ▲	Jet Proof	(IPX5)
⬤⬤	Immersion Proof	(IPX8)

Earthing and Bonding

E 1 Introduction

The subject of earthing and bonding in terms of installation power supplies and BS 7671 is not a highly complicated subject. BS 7671: 2008 comprises just nine pages of regulations on the subject, but there is probably more text written on this subject than any other.

This chapter focuses on BS 7671 requirements and solutions, and covers all that most readers will need. Supplementary information on earth electrodes is included in Appendix 12, as it is not relevant for many installations.

The interaction of the facets of an earthing system, which includes protective equipotential bonding, together with references to sections of this book, is represented in Figure E 1.1.

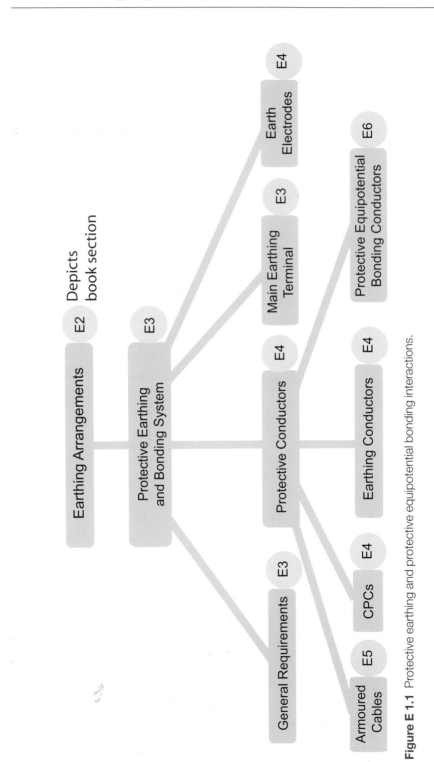

Figure E 1.1 Protective earthing and protective equipotential bonding interactions.

E 2   Earthing arrangements

BS 7671: 2008 defines or lists the types of system earthing in Part 2 – (definitions) and these are also defined in BS 7430: 1998.

As the system earthing arrangement is such an important component, the earthing arrangements are explained here.

The possible earthing arrangements and hierarchy are shown in Figure E 2.1.

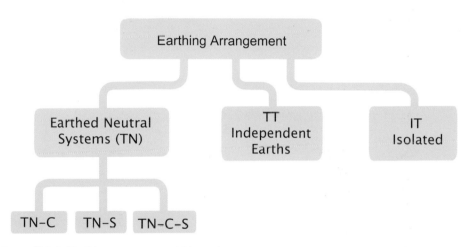

Figure E 2.1 Earthing arrangement hierarchy.

The system earthing arrangements are depicted by a code of letters. The first two explain the source and installation arrangement as explained in Figure E 2.2. Two additional letters classify a sub-arrangement of TN systems and relate to the combination of neutral and earth conductor. This is also detailed in Figure E 2.2.

The earthing arrangements are further explained with diagrams of both the supply and installation of typical connections, as shown in Figures E 2.3–E 2.8.

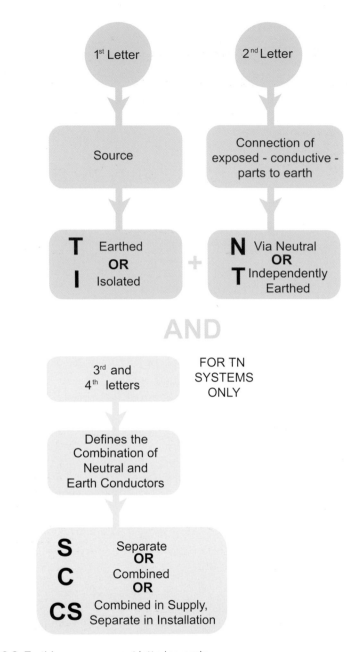

Figure E 2.2 Earthing arrangement lettering code.

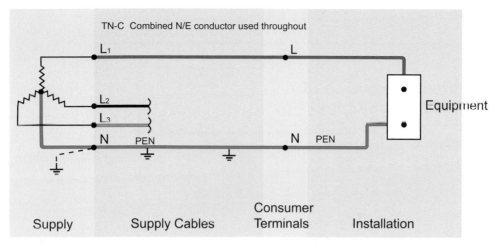

Figure E 2.3 TN-C system earthing.

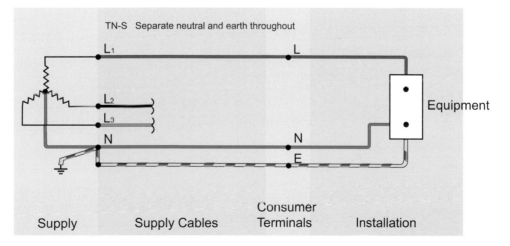

Figure E 2.4 TN-S system earthing.

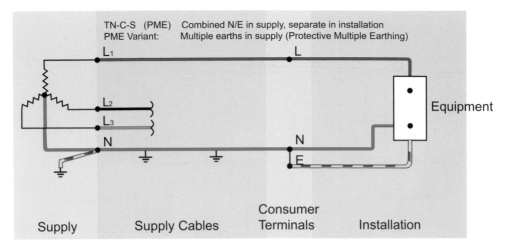

Figure E 2.5 TN-C-S system earthing with PME.

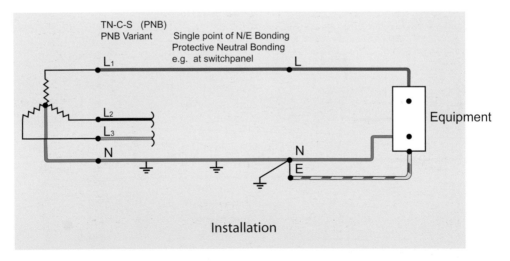

Figure E 2.6 TN-C-S system earthing with PNB.

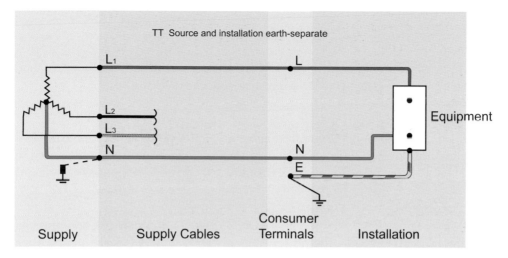

Figure E 2.7 TT system earthing.

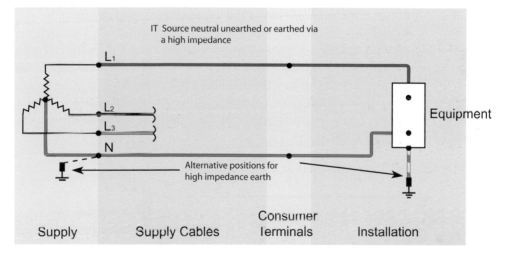

Figure E 2.8 IT system earthing.

Some relevant points to note about the various systems are now considered.

● TN-C

This system is only used in the UK by utilities and usually includes earthed sheath return wiring (ESRW) or earthed concentric wiring.

● TN-S

A true TN-S earthing system has the neutral-earth point at the source or right at the transformer. In practice this is rare and these systems are considered to be TN-C-S PNB (see Figures E 2.6 and E 2.9).

● TN-C-S (PME)

PME systems are used by the utility DNOs (distribution network operators) in most modern networks, mainly for economy and safety. A Government licence is required to own and operate this type of system.

● TN-C-S (PNB)

Figure E 2.6 is something that should be studied, as it is a variation of a TN-C-S system known as PNB, meaning protective neutral bonding. This variation of a TN-C-S system is not defined in BS 7671 but is defined in BS 7430: 1998. The electrical utility companies commonly use this PNB terminology, although it is not so common with electrical contractors or consulting engineers.

The TN-C-S PNB system is often used to describe the earthing arrangement in installations with on-site transformers; whilst some describe these arrangements as TN-S they are very close to the TN-C-S PNB system of Figure E 2.6, depending upon the position of the neutral earth link. A typical arrangement for an on-site transformer with the neutral earth link made in the main switchgear is shown in Figure E 2.9.

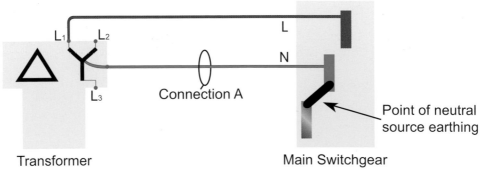

Figure E 2.9 Earthing arrangement for on-site substation.

'Connection A' in Figure E 2.9 should be considered as a combined neutral and earth conductor. This PNB arrangement is typical for most installations with an on-site transformer. The point of neutral system earthing is at one place within the main LV panel.

- TT

  In this system the source earth is independent of the installation earth, and hence for the latter there needs to be some form of earth electrode system.

- IT

  The source is either not earthed or earthed through a high impedance. A local IT 'system' can be created within a system but by considering Figure E 2.8, it can be seen that isolation of the local fabric is required and this can be difficult to achieve.

E 3 General requirements of earthing and bonding

The general requirements of BS 7671 for earthing are summarized in Table E 3.1.

Table E 3.1 General earthing requirements of BS 7671.

Requirement	Regulation
TN-S systems shall be directly connected to earthed point of source	542.1.2
TN-C-S systems shall be directly connected to earthed point of source via distributor's earth terminal	542.1.3
TT and IT systems shall be via earthing conductor to an earth electrode	542.1.4
Earth impedance shall be such that protective measure function is as designed	542.1.6 (i)
Conductors shall be adequately sized and robust	542.1.6 (ii) & (iii)
Different interconnected installations shall have adequately sized conductors for maximum fault current	542.1.8

Chapter 41 'Protection against electric shock', restructured and with new definitions for BS 7671: 2008, is described in Chapter C. The measures of complying with the requirements of Chapter 41 are called 'protective measures'. The protective measures for general application are:

- automatic disconnection of supply;
- double or reinforced installation;
- electrical separation;
- extra-low voltage provided by SELV or PELV.

The protective measure 'automatic disconnection of supply' is by far the most common and makes specific requirements for both protective earthing and protective bonding.

Again, whilst covered by Chapter C, we need to briefly review the requirements of the protective measure in order to establish earthing and bonding requirements. The basis of the measure is as follows.

- *Basic protection* (protection against direct contact) is protection from contact with live parts provided by basic insulation, or by barriers or enclosures.
- *Fault protection* (protection against indirect contact) is provided by protective earthing, protective equipotential bonding and automatic disconnection in case of fault.
  In the event of a fault between a live conductor and an exposed-conductive-part of equipment, sufficient fault current flows to operate (trip or fuse) the overcurrent protection.

For the protective measure to work, protective conductors are required to connect all exposed-conductive-parts (of equipment) to the earthing terminal. This is protective earthing and includes circuit protective conductors and the earthing conductor. To reduce shock voltages (touch voltages during a fault), and to provide protection in the event of an open circuit in the PEN conductor of a PME supply, protective equipotential bonding is also required. This requirement is fundamental to the automatic disconnection protective measure of Chapter 41.

A simple installation with both circuit protective conductors and protective bonding conductors is shown in Figure E 3.1.

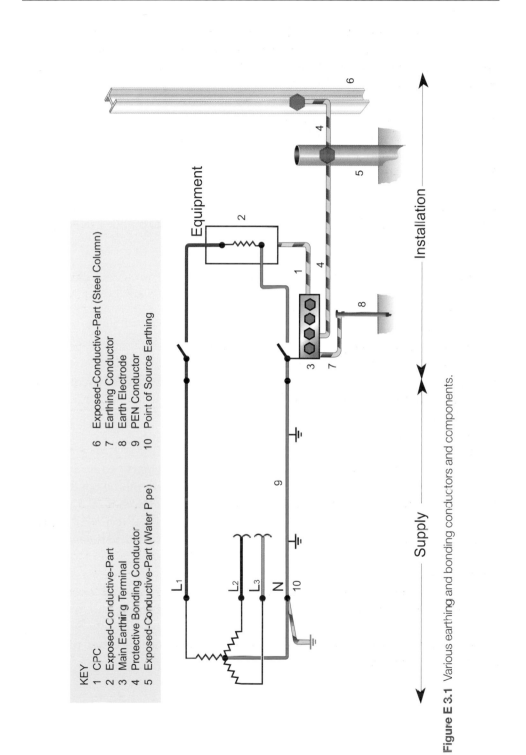

KEY
1 CPC
2 Exposed-Conductive-Part
3 Main Earthing Terminal
4 Protective Bonding Conductor
5 Exposed-Conductive-Part (Water Pipe)

6 Exposed-Conductive-Part (Steel Column)
7 Earthing Conductor
8 Earth Electrode
9 PEN Conductor
10 Point of Source Earthing

Equipment

Installation

Supply

Figure E 3.1 Various earthing and bonding conductors and components.

E 4   Protective conductors

E 4.1   General

Protective earthing requires exposed-conductive-parts (of electrical equipment, likely to become live in the event of a fault) to be connected (by protective conductors) to the main earthing terminal of the electrical installation. In the event of a fault, sufficient current will flow around the earth fault loop to cause the operation of the fault protective device (fuse, circuit breaker or RCD) and disconnect the fault.

Protective bonding conductors reduce the magnitude of voltages during fault conditions.

It should be remembered that the following are all types of protective conductor:

● circuit protective conductors (cpc);
● protective bonding conductors;
● earthing conductors.

It is important to establish and use the correct terminology; earthing conductors are often confused with cpcs. Earthing conductors are explained in E 4.5.

E 4.2   Physical types of protective conductor

BS 7671 recognizes that virtually any metallic conductive part can be used as a protective conductor, and this means as a cpc, a protective bonding conductor or an earthing conductor.

Table E 4.1 details metallic conductors and other metallic parts that can be used and provides guidance to their application.

It is strongly recommended that what could be described as 'natural' components be used as protective conductors. Often these elements exist as a matter of course, e.g. a metal pipe for another service. These items usually provide a much better solution than running extra green and yellow copper cable for protective conductors. The following solutions are recommended:

Table E 4.1 Recommended types of protective conductor.

Conductor	Recommended for use as cpc	Recommended for use as bonding conductor	Recommended for use as earthing conductor	Notes
Single core cable	✓	✓	✓	If less than 10 mm², shall be copper
Conductor in cable	✓	✓	✗	
Insulated or bare cable in enclosure with live conductors	✓	✓	✓	
Fixed insulated conductor	✓	✓	✓	cpc should be run in close proximity to line conductors. EMC reduction is better when run in enclosure or cable. If less than 10 mm² shall be copper
Fixed bare conductor	✓	✓	✓	
Cable sheaths or armouring	✓	✓	Possible	Possible with consideration
Metal conduit, trunking and ducting	✓	✓	✗	
Metal tray	Possible	✓	✗	Possible with consideration
Metal support systems	✗	✓ see note	✗	For use, continuity and size to be assured and precautions taken against removal
Structural metal	Possible	✓ see note	✓	
Metal water and service pipes, HVAC services and general metalwork	✗	✓ see note	✓	

- Always use metal conduit or trunking as a cpc where it is installed.
- If cable tray or ladder rack is used to support cables use it as a main bonding conductor.
- Utilize as much of other services and constructional material as possible for bonding.
- Utilize the structural steel for bonding.

Section E 6.3 provides more information on utilizing these solutions.

E 4.3   Sizing protective conductors

In order to prevent overheating of the protective conductor during a fault, the cross-sectional area of a protective conductor(s) shall be not less than that determined by the adiabatic formula as follows:

$$S = \frac{\sqrt{I^2 t}}{k}, \text{ or alternatively arranged as } S = \frac{I\sqrt{t}}{k}$$

where:

S is the nominal cross-sectional area of conductor in mm^2.

I is the value of fault current in amperes (rms for a.c.) for a fault of negligible impedance.

t is the operating time of the disconnecting device in seconds, corresponding to the fault current. It is found from the protective device characteristic curve.

k is a factor taking account of the resistivity, temperature coefficient and heat capacity of the conductor material, and the appropriate initial and final temperatures, see Tables 54.2 to 54.6 of BS 7671.

It should be noted that where the initial cable size has been adjusted following a thermal withstand check, further iterations may be necessary as the new size itself affects the prospective fault current.

Alternatively, the minimum cross-sectional area of a protective conductor can be determined by selection from Table 54.7 of BS 7671 (as below), but this virtually always yields a larger cable.

Table 54.7 of BS 7671 Minimum CSA of protective conductor in relation to the cross-sectional area of associated line conductor.

Cross-sectional area of line conductor S (mm^2)	If the protective conductor is of the same material as the line conductor (mm^2)	If the protective conductor is not the same material as the line conductor (mm^2)
$S \leq 16$	S	$k_1/k_2 \times S$
$16 \leq S < 35$	16	$k_1/k_2 \times 16$
$S > 35$	$S/2$	$k_1/k_2 \times S/2$

k_1 is the value of k for the line conductor, selected from Table 43.1 in Chapter 43 according to the materials of both conductor and insulation.
k_2 is the value of k for the protective conductor, selected from Tables 54.2 to 54.6 as applicable.

It should be noted that bonding conductors should not be sized using this method; it applies to cpcs and earthing conductors (see 543.1.1). For sizing of protective bonding conductors, see E6.

E 4.4   Protective conductors up to 16 mm²

The sizing of protective conductors depends upon only two factors: a size for disconnection times and a size for thermal capacity under fault conditions.

Instead of calculating the size of protective conductor for all combinations, the tables provided in this section give maximum earth fault loop impedance values for a number of protective device and protective conductor combinations.

When fuses to BS 1361, BS 88, or BS 3036 are used, this sets a maximum limit on the earth fault loop impedance. Tables E 4.2 to E 4.5 inclusive give maximum test earth fault loop impedances allowed for particular fuse and protective conductor

Table E 4.2 BS 3036 fuse maximum ELI values in ohms.

BS 3036 fuse maximum ELI values (Ω)					
Protective conductor size (mm²)	Fuse rating				
	5 A	15 A	20 A	30 A	45 A
1.0	7.7	2.1	1.4	NP	NP
1.5	7.7	2.1	1.4	0.9	NP
2.5	7.7	2.1	1.4	0.9	0.5
4.0 to 16	7.7	2.1	1.4	0.9	0.5

Notes:
NP: protective conductor/fuse combination not permitted.
A value of k of 115 from Table 54.3 of BS 7671 is used. This is suitable for PVC-insulated and sheathed cables to Table 5 of BS 6004.

Table E 4.3 BS 88 fuse maximum ELI values in ohms.

BS 88 fuse maximum ELI values (Ω)								
Protective conductor size (mm²)	Fuse rating							
	6 A	10 A	16 A	20 A	25 A	32 A	40 A	50 A
1.0	6.87	4.12	2.18	1.43	1.16	0.66	NP	NP
1.5	6.87	4.12	2.18	1.43	1.16	0.84	0.64	NP
2.5 to 16	6.87	4.12	2.18	1.43	1.16	0.84	0.66	0.49

Notes:
NP indicates protective conductor/fuse combination not permitted.
A value of k of 115 from Table 54.3 of BS 7671 is used. This is suitable for PVC-insulated and sheathed cables to Table 5 of BS 6004.

Table E 4.4 BS 1361 fuse maximum ELI values in ohms.

BS 1361 fuse maximum ELI values (Ω)					
Protective conductor size (mm²)	Fuse rating				
	5 A	15 A	20 A	30 A	45 A
1.0	8.43	2.65	1.37	0.77	NP
1.5	8.43	2.65	1.37	0.93	NP
2.5	8.43	2.65	1.37	0.93	0.51
4.0	8.43	2.65	1.37	0.93	0.68
6.0 to 16	8.43	2.65	1.37	0.93	0.77

Notes:
NP indicates protective conductor/fuse combination not permitted.
A value of k of 115 from Table 54.3 of BS 7671 is used. This is suitable for PVC-insulated and sheathed cables to Table 5 of BS 6004.

combinations. When the protective conductor is small the limiting factor is the adiabatic equation. For larger sizes of protective conductor the limiting factor is the shock protection requirement.

When circuit breakers are used to protect the circuit, because shock protection requirements require that circuit breakers operate instantaneously in the event of a fault, the most practical approach is to set a minimum protective conductor size for each type and rating of circuit breaker. For completeness in Table E 4.5 below, the maximum test loop impedance is included and these values ensure instantaneous operation (0.1 s) of the circuit breaker.

Table E 4.5 Cable size and maximum ELI values for BS EN 60898 circuit breakers.

Circuit breaker type		Device rating A										
		6	10	16	20	25	32	40	50	63	80	100
B	Min cable size (mm²)	1	1	1	1	1	1	1	1	1.5	1.5	1.5
	Max ELI (Ω)	6.2	3.7	2.3	1.9	1.5	1.2	0.93	0.74	0.59	0.46	0.37
C	Min cable size (mm²)	1	1	1	1	1	1	1	1.5	1.5	2.5	2.5
	Max ELI (Ω)	3.1	1.9	1.2	0.93	0.71	0.58	0.46	0.37	0.29	0.23	0.18
D	Min cable size (mm²)	1	1	1.5	1.5	2.5	2.5	4	4	6	6	10
	Max ELI (Ω)	1.6	0.93	0.58	0.46	0.37	0.29	0.23	0.18	0.15	0.12	0.09

E 4.5 Earthing conductor

The earthing conductor connects the system earth bar or main earthing terminal (MET) to the system means of earthing (see Figures E 3.1 and E 6.5).

It is sized as for other protective conductors, namely either by the adiabatic equation:

$$S = \frac{I\sqrt{t}}{k}$$

or by using Table 54.7 of BS 7671; this is explained in Section E 4.3.

For PME installations the size of the earthing conductor must be no less than the sizes in Table E 4.6.

Table E 4.6 PME (TN-C-S) minimum earthing conductor sizes.

PME (TN-C-S) main protective bonding conductor sizes (mm²)								
	Supply neutral conductor size (mm²)							
	25	35	50	70	95	120	150	Over 150
Earthing conductor min. size (mm²)	10	10	16	25	25	35	35	50

Note: Electricity utility distributors may require a minimum size of earthing conductor at the origin of the supply of 16 mm² copper or greater for TN-C-S (PME) supplies.

E 5 Armoured cables as protective conductors

E 5.1 General

It has been common practice in the United Kingdom to use the armouring of steel wire armoured cables as the protective conductor. This practice complies with BS 7671 and is most desirable for a number of reasons. Armouring provides a low-impedance, robust, protective earthing system and will greatly assist any potential EMC issues compared with the practice of running external cpcs.

Checking that the armouring is of sufficient cross-sectional area is complicated by the armouring being of a different material (steel) to that of the conductor (copper or aluminium). It is also complicated by the cross-sectional area for some two- and three-core cables being less than that given by Table 54.7. In the past, various organizations have suggested that certain cable configurations and sizes are insufficient for compliance with BS 7671.

However, the remedy of the suggested undersized armour was to run an external cable cpc in the vicinity of the cable. This subject itself posed unanswered problems of the current sharing between the armour and the external cpc. It was thought that current sharing under fault conditions would not be in line with the respective impedances due to the magnetic field interaction of the armour. There was no recognized method of approximating this current sharing.

Due to this, during 2007 ECA sponsored research on this subject by the cables division of the Electrical Research Association (ERA). The report is included in Appendix 16 and its conclusions are given below.

E 5.2 ERA report on current sharing between armouring and cpc

Conclusions

Prior to carrying out the test work described above it was thought that the proportion of any earth fault current carried by an external cpc, run in parallel with the armour of a cable, would be less than that predicted from the ratio of the armour and cpc resistance. The test work has shown that this is not the case and that using the resistance ratio will give a reasonable estimate of the current carried by the external cpc. Comparison of the fault current withstand of the armour of cables to BS 5467, using the k values given in Chapter 54 of BS 7671, has shown that for all cables except the 120 mm^2 and 400 mm^2 2-core cables the fault current withstand of the armour is greater than the fault current required to operate a BS 88 fuse. Thus a supplementary external cpc is generally not required to increase the fault current withstand of an armoured cable.

A supplementary external cpc would only be required if the earth fault loop impedance needed to be reduced to meet the values tabulated in Chapter 41 of BS 7671. This could be the case for a long cable run where protection against indirect contact was achieved by automatic disconnection of the supply and the use of a residual current device was not appropriate.

(Continued.)

(Continued.)

Empirical equations have been derived for calculating the prospective fault current where an external cpc is installed in parallel with steel wire armour. Empirical equations have also been derived for calculating the current sharing between the armour and the external cpc. Calculation of the current sharing between the armour and an external cpc has shown that if a small external cpc is run in parallel with the armour of a large cable there is a risk that the fault current withstand of the external cpc will be exceeded. Because of this it is recommended that the cross-sectional area of the external cpc should not be less than a quarter of that of the line conductor.

E 5.3 ECA advice and recommendations

The ECA recommends the following:

- Use all armoured cables as a cpc without the need for a thermal withstand check, with the exception of 400 mm^2 2-core.
- Note: The 120 mm^2 cable was within a few per cent of the size required (required I^2t of 33 062 500, actual 33 800 000 after 5 seconds) and should be used without calculation unless it is intended to run the cable at full current-carrying capacity. It is very difficult to load a cable of this size at its full current-carrying capacity for a continuously long period.
- Where calculation software specifies external cpcs for ELI reduction, use a cpc size of at least 25% of the line conductor csa (perhaps after manually checking calculations).
- Manually check ELI calculations where software specifies small external cable cpcs.

E 6 Protective equipotential bonding

E 6.1 Purpose of protective equipotential bonding

This section discusses main equipotential bonding. The purpose of protective equipotential bonding is to:

- reduce shock voltages in the event of a fault; and
- reinforce the connection with the earth of the installation.

E 6.2   BS 7671 requirements

In every installation, protective bonding conductors are required to connect the protective earthing terminal to extraneous-conductive-parts including to the earthing system (via the main earthing terminal). Chapter 41 has a different list of suggested items to bond than does Chapter 54 but both should be used. The comprehensive list is as follows:

- circuit protective conductors;
- water installation pipes;
- gas installation pipes;
- other installation pipework and ducting;
- exposed, metallic structural parts of the building;
- functional earthing conductors where used;
- lightning protection system (LPS) in accordance with BS EN 62305[1].

[1] At the time of drafting, BS EN 62305 was under modification to state that this bond was the responsibility of the LPS contractor in terms of both sizing and installation.

E 6.3   Bonding solutions for the modern installation

The question still remains as to how to carry out bonding in a modern installation. Historically, after publication of the 15th Edition of the IEE Wiring Regulations in 1981, the industry went through what can only be described as a ridiculous period of practice. It was common to bond all exposed metalwork, including such items as metal window frames, individual metal floor or ceiling tiles and cross bonding to virtually everything. If consulting engineers did not insist on the practice, then the installation contractor did!

The modern trend is now well away from this, but the question remains of what to do about continuity of, say, the structural steel, or air conditioning ductwork systems.

The solution advocated by the ECA for many years now is provided in this section.

Commercial and industrial installations

● Take one bonding conductor to each of the items in the list given in Section E 6.2.

That's it! See Figures E 6.1 to E 6.4 and accompanying notes.

Using constructional elements for bonding

Figure E 6.1 shows a typical commercial or industrial structure. There is a metal superstructure either with steel beams and columns, or steel reinforced concrete. There are many interconnected star and parallel metallic and conductive parts. Many of these are not shown on the diagram, but consider the amount of services and their support metalwork for the ductwork, water pipework, gas piping, external cladding, false floor, false ceiling, lifts and, of course, the electrical services. There really is no need for any green and yellow cable at all. Continuity tests on this metalwork virtually always indicate a continuity approaching zero ohms.

The best solution for main protective equipotential bonding is that indicated in Figure E 6.4, but often the solutions shown in Figures E 6.1 or E 6.2 are installed. The figures now follow.

Figure E 6.1 shows a typical solution which involves running separate bonding conductors to the extraneous-conductive-parts. This is unnecessary.

Figure E6.2 shows a typical solution which involves running separate bonding conductors to the extraneous-conductive-parts but now installed on metal cable tray or ladder rack. This is often undertaken whether or not the tray or ladder rack was installed for power cables.

Figure E 6.3 shows a much better and more economic solution, where the cable tray or ladder rack has been used as a protective conductor.

Figure E 6.4 shows the best solution for most installations of this nature. The building structure has been used as a protective conductor. The cable tray is not shown on this diagram as it may have been installed for just a few cables, which have now been clipped directly to the structure. Unfortunately, this technique is under-utilized, mainly due to custom and practice. If the diagrams are considered again, it is obvious that the structure and metal services will form part of the bonding network whether you like it or not.

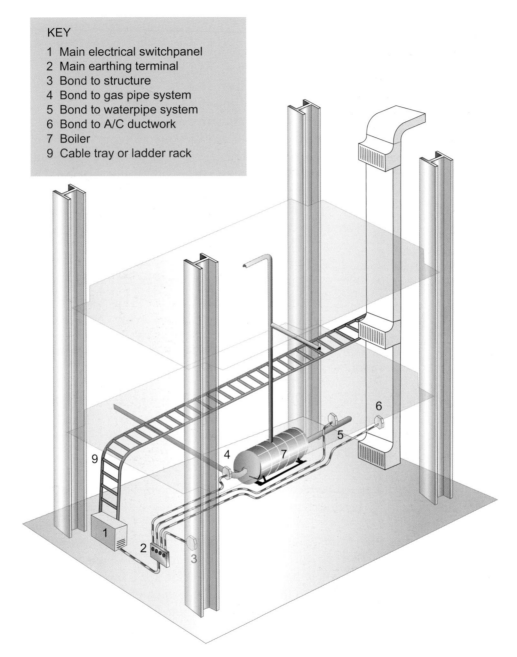

KEY

1 Main electrical switchpanel
2 Main earthing terminal
3 Bond to structure
4 Bond to gas pipe system
5 Bond to waterpipe system
6 Bond to A/C ductwork
7 Boiler
9 Cable tray or ladder rack

Figure E 6.1 Protective equipotential bonding – via multiple cables.

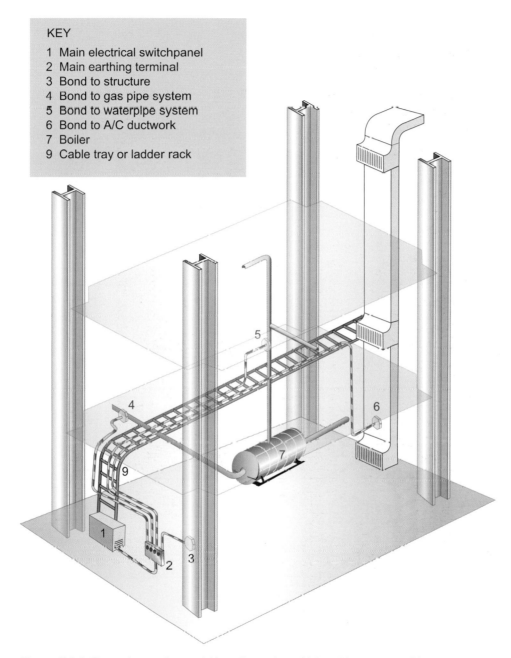

KEY

1 Main electrical switchpanel
2 Main earthing terminal
3 Bond to structure
4 Bond to gas pipe system
5 Bond to waterplpe system
6 Bond to A/C ductwork
7 Boiler
9 Cable tray or ladder rack

Figure E 6.2 Protective equipotential bonding – via multiple cables run on cable tray.

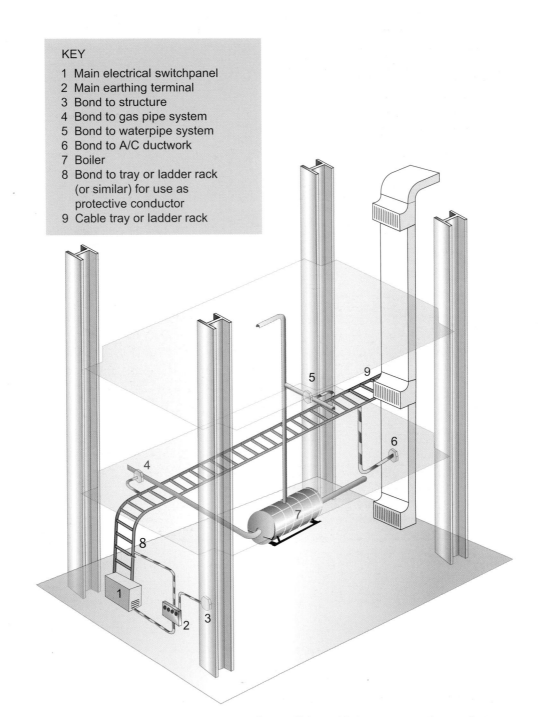

KEY
1 Main electrical switchpanel
2 Main earthing terminal
3 Bond to structure
4 Bond to gas pipe system
5 Bond to waterpipe system
6 Bond to A/C ductwork
7 Boiler
8 Bond to tray or ladder rack
   (or similar) for use as
   protective conductor
9 Cable tray or ladder rack

Figure E 6.3 Protective equipotential bonding – utilizing cable tray as protective conductor.

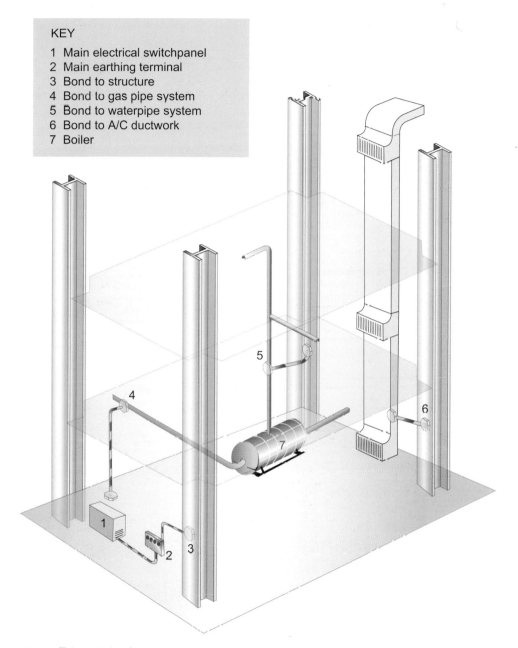

KEY

1 Main electrical switchpanel
2 Main earthing terminal
3 Bond to structure
4 Bond to gas pipe system
5 Bond to waterpipe system
6 Bond to A/C ductwork
7 Boiler

Figure E 6.4 Protective equipotential bonding – utilizing structure as protective conductor.

Domestic installations

● Take one bond to incoming metallic water and gas pipes.
● If there is a metallically piped central heating system, bond it in one location. This is not necessary where any metallic pipe, bonded elsewhere (e.g. metallic water pipe), is connected and tests satisfactorily for continuity.
● If the structure is metallic, take a bond to one location; this is not necessary for intermediate metallic stud partitions or individual metal beams and the like.
● The central heating system will probably not need a cable bond connection if connected by metallic gas or water pipes which have themselves been bonded (see Figure E 6.5).

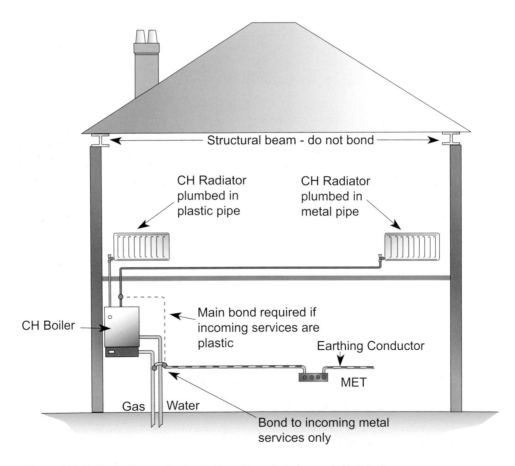

Figure E 6.5 Protective equipotential bonding – in a domestic installation.

Figure E 6.5 shows a typical domestic dwelling, and main protective bonding requirements.

The radiator served via plastic pipe does not require a bond. The radiator served via metal pipes will need a bond if the boiler is not itself bonded. However, it would be better practice to bond the boiler or a pipe close to the boiler.

E 6.4 Sizing protective bonding conductors

The sizing of bonding conductors can get confusing when reading Chapter 54 of BS 7671. This is due to the fact that the suggested minimum sizes relate to the size of the line and earthing conductor for TN-S supplies, and the supply neutral conductor for TN-C-S (PME) supplies.

Tables E 6.1 and E 6.2 provide an easy look up for main protective bonding conductor sizes.

Table E 6.1 TN-S, PNB, TT main protective bonding conductor sizes.

	Line conductor size (mm²)												
	25	35	50	70	95	120	150	185	240	300	400	500	600
Earthing conductor min. size (mm²)	16	16	25	35	50	70	95	95	120	150	240	300	300
Bonding conductor min. size (mm²)	10	10	16	16	25	25	25	25	25	25	25	25	25

Notes:
Smaller sizes are possible by using $S = (I\sqrt{t})/k$ to calculate earthing conductor size; main bonding is half the size of this up to a maximum of 25 mm². A minimum conductor size of 6 mm² shall be used and conductors of 10 mm² or less shall be copper.
For conductors of other material, the material must be sized to achieve the equivalent conductance of the copper of the relevant size. See Appendices for data.
IT system earthing conductors may be sized using this table.

Table E 6.2 PME (TN C S) main protective bonding conductor sizes.

	Supply neutral conductor size – mm²							
	25	35	50	70	95	120	150	Over 150
Earthing conductor min. size (mm²)	10	10	16	25	25	35	35	50
Bonding conductor min. size (mm²)	10	10	16	25	25	35	35	50

Note: Electricity utility distributors may require a minimum size of earthing conductor at the origin of the supply of 16 mm² copper or greater for TN-C-S (PME) supplies.

E 6.5   Domestic protective equipotential bonding layouts

This section provides diagrams of typical protective equipotential bonding layouts. It has been added to provide further guidance on typical layouts of bonding conductors, and also assists with recognizing earthing arrangements. Many other layouts are, of course, possible.

Figure E 6.6 shows the main earthing terminal connected via the earthing conductor to the utility cut-out stud terminal (TN-C-S).

Figure E 6.7 shows the main earthing terminal connected via the earthing conductor to the sheath of the utility supply cable (TN-S).

Figure E 6.8 shows the main earthing terminal connected via the earthing conductor to the installation earth electrode (TT). The earthing conductor in a TT installation may only need to be $6\,mm^2$ although it is shown as $16\,mm^2$ in this diagram.

E 6.6   Supplementary equipotential bonding

Chapter 41 of BS 7671: 2008 lists two additional measures intended to supplement the main protective measures for protection against electric shock. The additional measures are RDCs and supplementary equipotential bonding. The subject of RCDs is covered in this book in Chapters C and D.

Supplementary bonding is additional protection to fault protection, and may be required where the disconnection times of Table 41.1 of BS 7671 cannot be achieved. However, the use of supplementary bonding does not exclude the need to disconnect the supply for other reasons, for example, for protection against overcurrent.

Supplementary bonding may also be required for special locations such as:

- locations with a bath or shower 701.415.2;
- swimming pools and other basins 702 .411.3.3;
- locations with livestock 705.415.2.1;
- conducting locations with restricted movement 706.410.3.10 (iii).

It should be noted that there is no requirement in BS 7671 to install supplementary bonding in kitchens.

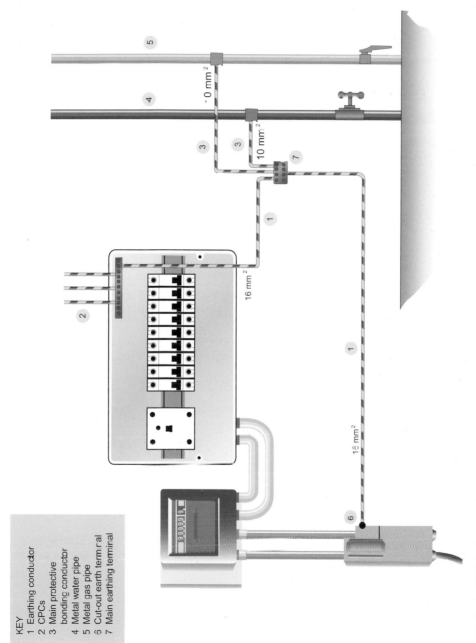

KEY
1 Earthing conductor
2 CPCs
3 Main protective
   bonding conductor
4 Metal water pipe
5 Metal gas pipe
6 Cut-out earth terminal
7 Main earthing terminal

Figure E 6.6 Typical TN-C-S protective equipotential bonding layout.

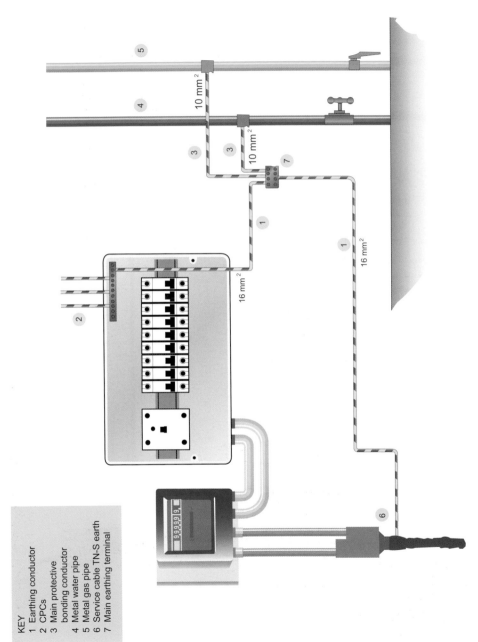

KEY
1 Earthing conductor
2 CPCs
3 Main protective
  bonding conductor
4 Metal water pipe
5 Metal gas pipe
6 Service cable TN-S earth
7 Main earthing terminal

Figure E 6.7 Typical TN-S protective equipotential bonding layout.

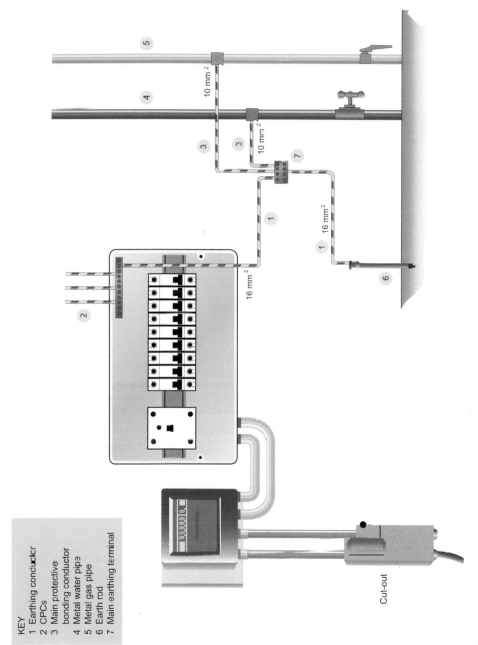

KEY
1 Earthing conductor
2 CPCs
3 Main protective
  bonding conductor
4 Metal water pipe
5 Metal gas pipe
6 Earth rod
7 Main earthing terminal

Cut-out

Figure E 6.8 Typical TT protective equipotential bonding layout.

E

The sizes of supplementary bonding conductors are summarized in Table E 6.3 for cables not mechanically protected, which is the usual case. For completeness, Table E 6.4 is for cables which are mechanically protected, i.e. run in conduit, trunking or similar.

Table E 6.3 Supplementary bonding conductor sizes – unprotected cables.

Circuit conductor size (mm²)	Exposed to exposed-conductive-parts (mm²)	Exposed to extraneous-conductive-parts (mm²)	Extraneous to extraneous-conductive-parts (mm²)
1.0	4	4	4
1.5	4	4	4
2.5	4	4	4
4	4	4	4
6	6	4	4
10	10	6	4
16	16	10	4

Note: If one of the extraneous-conductive-parts is connected to an exposed-conductive-part, the bond must be no smaller than that required for bonds between exposed-conductive-parts.

Table E 6.4 Supplementary bonding conductor sizes – protected cables.

Circuit conductor size (mm²)	Exposed to exposed-conductive-parts (mm²)	Exposed to extraneous-conductive-parts (mm²)	Extraneous to extraneous-conductive-parts (mm²)
1	1	1	2.5
1.5	1	1	2.5
2.5	2.5	1.5	2.5
4	4	2.5	2.5
6	6	4	2.5
10	10	6	2.5
16	16	10	2.5

Note: If one of the extraneous-conductive-parts is connected to an exposed-conductive-part, the bond must be no smaller than that required for bonds between exposed-conductive-parts.

E 7 High earth leakage installations

The subject of circuits or equipment with high earth leakage currents was covered in a Special Installations section (607) in previous versions of BS 7671. This section was deleted in the international IEC document and for BS 7671: 2008 the requirements are contained within Chapter 54, Section 543.7.

The designer of the installation needs to assess whether the equipment circuit is likely to carry earth leakage current. It is recommended to use a figure of 1 mA earth leakage per workstation outlet for personal computers.

The requirements are summarized in Table E 7.1 for circuits and Table E 7.2 for individual items of equipment. The option of using earth monitoring has not been included as it is a rather specialist area and very rarely used.

Table E 7.1 Requirements for circuits with collective leakage exceeding 10 mA.

Requirement	Regulation	Comments
Circuits shall have dual cpc terminated separately at supply and outlets, or	543.7.1.3	Applies to ring socket-outlet circuits
Circuits shall have a 4 mm² copper cpc, or	543.7.1.3	
Circuits shall have a 10 mm² cpc	543.7.1.3	
'Spur'- off rings to be duplicated	543.7.2.1	

Table E 7.2 Requirements for individual items of equipment with leakage exceeding 3.5 mA.

Requirement	Regulation
Items of equipment with leakage between 3.5 mA and 10 mA: hard wire or connect via 'industrial' plug and socket to BS EN 60309–2	543.7.1.1
Items of equipment with leakage of 10 mA and above: hard wire or connect via 'industrial' plug and socket to BS EN 60309–2 with cpc of 2.5 mm², or 4 mm² if >16 A	543.7.1.2

Inspection, Testing and Certification (Part 6)

F 1   Introduction

Part 6 of BS 7671: 2008 is entitled 'Inspection and Testing' but covers the subjects of certification and reporting as well as inspection and testing.

The Part comprises three chapters as follows;

- 61 Initial Verification;
- 62 Periodic Inspection and Testing;
- 63 Certification and Reporting.

The subject of Periodic Inspection and Testing is discussed in Appendix 15 of this book as many organizations do not undertake such work, which is generally regarded as a little more 'specialist'.

Chapter F therefore discusses and provides explanation and guidance on the inspection, testing and certification of new installations, including alterations and additions.

The main part of the chapter addresses the 'initial verification' of the Regulations and confirmation that the installation meets the requirements of the Regulations.

F 1.1   Inspection and testing – an integrated procedure

The activities of carrying out visual inspection and of then carrying out testing should be considered as complementary procedures. These procedures should not be considered to be separate functions carried out by separate individuals or organizations; this is particularly true of the inspection confirmation part of the inspection and testing.

To illustrate this point: in general there would be little point in carrying out a continuity test on a cable if it had been found that some of its connection terminals were loose. The defect would need to be remedied before the test was conducted.

Another aspect to consider are the stages of inspection and stages of testing for larger installations. For many such installations the inspection is formally carried out at the end of the constructional element of the electrical installation process; often a separate 'test' electrician or engineer undertakes this task.

However, the amount of visual inspection undertaken during the electrical construction stage is often very significant. Most electricians check visual elements like polarity, tightness of connections, and indeed many of the items required under the visual inspection element of inspection and testing (see Section F 2 below). These visual checks are done as a matter of course, but often the 'test engineer' completes the visual inspection separately and often without consultation with the installation electricians. It is suggested that this 'inspection' facet of the inspection and testing process is more efficient, and generally better, if carried out 'inherently' by the installers. Of course, there are exceptions; a particular example is where semi-skilled labour is being utilized for installation.

It is suggested, for larger organizations, that a system be created that allows the installer to confirm facts about the visual inspection. Table F 2.1 suggests items that can and should be carried out during the electrical construction procedure.

F 2　Visual inspection

The visual inspection part of the initial verification is the process of assessing the installation prior to testing. Regulation 611.2 specifies the purpose of the visual inspection, summarized as follows:

- equipment complies with a product standard (see Section D 2);
- equipment is correctly selected and erected (see Chapter D);
- equipment is not damaged.

Table F 2.1 summarizes the requirements for the inspection part of the initial verification. To be consistent with the restructured Chapter 41 (see Chapter C) the protective measures like 'placing out of reach', and similar, have been removed from this table, as they are not considered at the installation stage. The table has been constructed with a 'should have an automatic tick' column. This column indicates facets that should be automatically inspected by the installer. For example, how can an installer install a cable if he docs not know the designer's specified cable size? As a further example, the installer should not be terminating cables if he does not know BS 7671 aspects relating to the tightness of connections; thus, he should be inspecting this part of the work. This column should be inherently carried out by the installing operative, completed on the inspection checklist paperwork and passed to the 'test engineer' if this is a separate individual.

Table F 2.1 Visual inspection requirements of BS 7671: 2008.

Inspection	Should have automatic tick	At 'installed' stage	Confirmed separately at final 'testing' stage	Notes
Erection methods	✓	✓	Not required	
Conductor terminations	✓	✓	Not required	
Identification (colour/labelling of cables)	✓	✓	Not required	
Installation of cables (mechanical protection, 522 etc.)	✓	✓	Not required	
Cable size as design specification	✓	✓	Not required	
Building sealing around cables and trunking etc.		✓	Not required	Spot checks may be useful
Presence of means of earthing	✓	✓	Not required	
Presence of protective conductors, cpcs and bonding	✓	✓	Not required	
Isolation	✓	✓	Not required	
Warning notices	✓	✓	Not required	
Diagrams	✓	✓		
Equipment to standards, and IP	✓	✓	Not required	Spot checks may be useful
Detrimental influences		✓	Spot checks may be useful	
Adequate access to switchgear	✓	✓	Not required	Spot checks may be useful
Overcurrent devices correctly sized as specified	✓	✓	Not required	Spot checks may be useful
Presence of RCD protection where required	✓	✓	Not required	Spot checks may be useful
Undervoltage devices where required		✓	Check	

The detail of inspection to this table is covered in the other chapters of this book, and is not expanded upon here with the exception of some labelling and warning notices; for these, refer to BS 7671: 2008 directly.

It should be apparent that if an organization employs separate 'inspection and test engineers', much of the inspection work should be a conversation and completing paperwork exercise with the test engineer. This exercise is better if the inspection paperwork originates with the designer and is passed on to the test engineer. In these situations it should not be the function of the 'test engineer' to repeat inspections of items confirmed by the installers; this is not a requirement of the Regulations.

F 3 Testing

F 3.1 Introduction – pass and fail nature

This section discusses methods for achieving the physical testing requirements of BS 7671. It should be noted that, in some books and advice, too much emphasis is placed on particulars of test methods, often in areas that are not important and do not make any difference to the safety of an installation.

It should be remembered that, of the seven 'everyday' tests required by BS 7671, five are really pass/fail tests. Also, the test methods suggested by industry guidance and manufacturers' literature are just that – guidance. Other methods can and will produce satisfactory results and may be used; and this is, of course, true of this part of this book.

F 3.2 Required tests

The tests required by Part 7 of BS 7671: 2008 fall into two categories:

- The seven 'everyday' tests. These are regularly and continually used by installers and are described in this section of the book.
- Additional tests that may be used occasionally. They would include, for example, testing the insulation of an IT 'floor structure'. These are discussed in Appendix 14.

The 'everyday tests' required by Part 6 of BS 7671: 2008 have been required in the Regulations for a considerable number of years. The Regulations and guidance

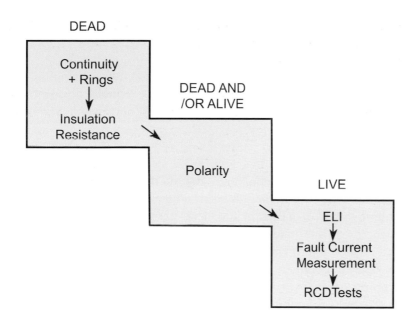

Figure F 3.1 'Everyday tests' order and state of energization.

have suggested an order of tests based on safety and logic. The suggested order is to carry out at least two tests prior to energization (cpc continuity and insulation resistance, with the possible addition of polarity checks) and the remainder to be carried out after energization. A diagram of the seven 'everyday tests' and the suggested order of tests is given in Figure F 3.1.

Working to the suggested order of tests is recommended, but you should bear in mind that some tests will require disconnection of the circuit's neutral; the exact order should not be cast in stone, although the dead tests should certainly be completed prior to energization for safety reasons.

F 3.3   Continuity testing

This is the first suggested test as it is important for the safety of the circuit, and it helps confirm a reference for the remainder of the tests. Continuity testing is carried out on all protective conductors including cpcs, main and supplementary protective bonding conductors.

The test is carried out with a low d.c. voltage continuity tester, and this may detect loose and unsound connections; other instruments may be used.

It is important to establish the path of the conductor that is being measured, as often there will be parallel paths in circuit. This is virtually always the case in commercial and industrial installations where parallel metalwork has a significant bearing on readings.

Consider the luminaire circuit in Figure F 3.2 installed in a typical commercial installation. The figure shows a luminaire mounted with a substantial metal fixing to the structure. There are numerous earth return paths back to the source earth point including the cable cpc, trunking, metal supports, the cable ladder rack and the superstructure of the building. It is unrealistic to disconnect the luminaire from these to test for continuity. Even if the luminaire had an 'insulated' cable supply, unlikely in this environment, the cpc would need to be disconnected at every luminaire on the circuit in order that the cable cpc be tested correctly. This would defeat the object, as the connections are the most vulnerable parts of the circuit. Thus, the only test possible is to test for continuity of the luminaire, including the multiple parallel return paths.

In effect, the reading is confirming that the equipment under test is earthed and, with this test, it is possible that a cable cpc is open circuit. Some do not like this situation and it is one of the reasons that the ECA would recommend using metallic trunking or metallic conduit as a cpc and not a cable cpc within these systems.

This brings us to the two popular methods for continuity testing, namely:

1 'wandering lead' method;
2 utilization of circuit cable by 'shorting'.

Wandering lead method

The author, mainly for the reasons given above concerning parallel returns, prefers this method. Figure F 3.2 shows a 'wandering lead' continuity test.

You will notice by studying the figure that the cpc has not been removed from the earth terminal at the distribution board. This is due to the fact that there are parallel metallic earth return paths and removal of this conductor will not change this.

Once set up, testing using the 'wandering lead' method is quick and simple. Metalwork in the vicinity of the equipment under test can be quickly checked (often all that is needed is a piece of timber with cable and spike!). Extraneous-conductive-parts can be checked using the same method.

Continuity of CPC by 'Wandering Lead'

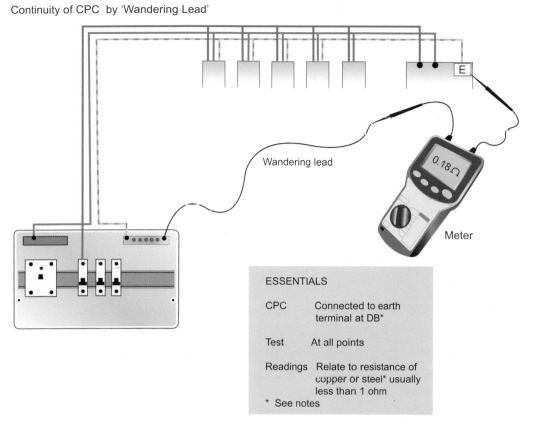

ESSENTIALS

CPC   Connected to earth
          terminal at DB*

Test   At all points

Readings Relate to resistance of
          copper or steel* usually
          less than 1 ohm
* See notes

Figure F 3.2 Continuity of cpc by 'wandering lead' method.

It is recommended that this method be used for commercial installations with parallel earth return paths. Also, the recording of a resistance value for each circuit is rather pointless, as parallel paths do not fall neatly into 'circuits'. Instead, all equipment relevant to a distribution board is tested and the following recorded on the test sheet relating to the distribution board:

'All circuits tested for continuity maximum reading 1.2 ohms (luminaire director's office)'

As for readings, many tests using this method will show readings approaching zero or typically less than one or two ohms.

Utilizing the circuit cable and $R_1 + R_2$ method

The second method involves using the circuit cable and shorting or linking it out at one end as shown in Figure F 3.3.

Continuity using circuit cable

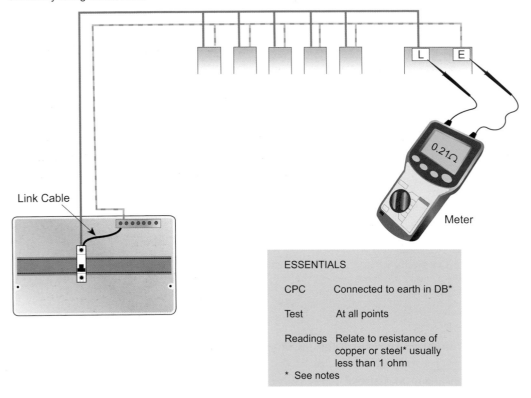

Figure F 3.3 Continuity of cpc by circuit cable method.

ESSENTIALS

CPC	Connected to earth in DB*
Test	At all points
Readings	Relate to resistance of copper or steel* usually less than 1 ohm

* See notes

The method is simple enough, and if the line conductor is used, this method can be used to record the circuit's $R_1 + R_2$ value. However, the comments made in the 'wandering lead' method about parallel metallic return paths also apply to this method, and therefore in virtually all commercial and industrial applications this test only measures R_1 plus a value of a parallel multi-connected return path. It is therefore a little misleading to record results for this method as a circuit's $R_1 + R_2$ value without a cautionary note on the test sheet.

When compared with the 'wandering lead' method, this test method is slow to repeat at all outlets on a particular circuit (except for socket outlets) as it involves access to equipment terminals.

The method does find favour with some, and can be used to establish a circuit's earth fault loop impedance value as discussed later.

F 3.4 Ring continuity

The continuity of ring final circuits is something that is a little more involved. Methods have evolved over the years to establish whether the rings are indeed wired as ring circuits, and not 'shorted' forming a 'figure of eight' layout and similar.

Whether the 17th Edition requires this test is questionable. The best advice here is not to use ring circuits for commercial and industrial installations.

For completeness and where ring final circuits are used, the following continuity tests are recommended. It should be noted that the stage 3 tests involving using the circuits cpc will suffer from the problem of parallel multi-metallic return paths as mentioned in Section F 3.3, and unless circuit wiring, socket outlets and accessory boxes are insulated from earth this test stage should not be used.

STAGE 1

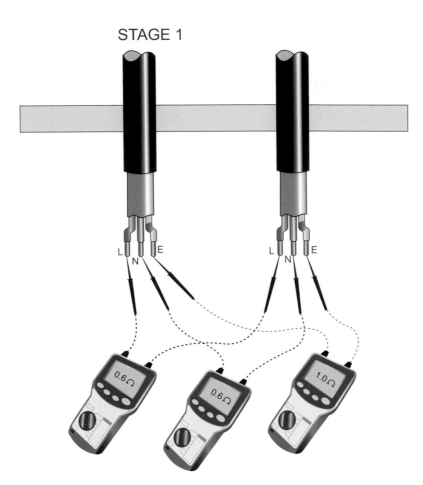

Figure F 3.4a Continuity of ring final circuits, stage 1.

Stage 1 – open loop resistances

Measure the end-to-end resistance of each conductor, R_1 (line), R_n (neutral) and R_2 (cpc) respectively as indicated in Figure F 3.4a. An open circuit result would suggest either the incorrect conductors have been selected or the circuit is incorrectly terminated. As phase and neutral conductors will be of the same cross-sectional area, the resistance values obtained for both conductors should be similar.

Where twin and earth cables to BS 6004 are used, and the cross-sectional area of the cpc is reduced in comparison with the live conductors, the resistance of R_2 will be comparatively higher than that of the phase and neutral loops.

STAGE 2

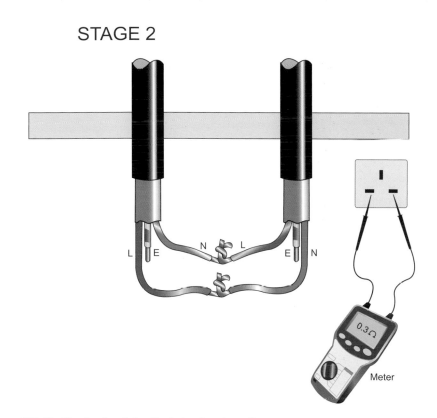

Figure F 3.4b Continuity of ring final circuits, stage 2.

Stage 2 – interconnected L-N

With phase and neutral interconnected as Figure F 3.4b, measure resistance between phase and neutral conductors at each socket outlet using a continuity tester or similar instrument. If the ring is not interconnected the measurements taken on the ring circuit will be similar. The measurements obtained will be approximately one quarter of the resistance of the sum of the open loop resistances from stage 1, that is $(R_1 + R_n)/4$.

STAGE 3

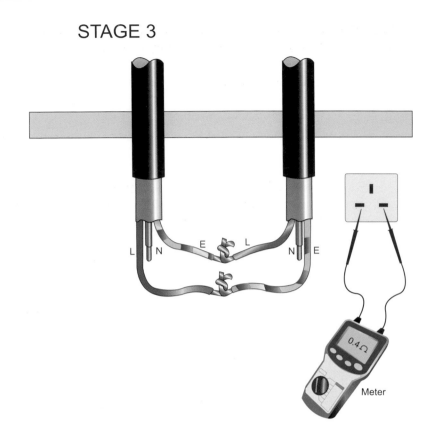

Figure F 3.4c Continuity of ring final circuits, stage 3.

Stage 3 – interconnected L-cpc (for all insulated systems)

With phase and cpc interconnected as Figure F 3.4c, measure resistance between phase and cpc conductors at each socket outlet using a continuity tester or similar instrument. If the ring is not interconnected the measurements taken on the ring circuit will be similar. The measurements obtained will be approximately one quarter of the resistance of the sum of the open loop resistances from stage 1, that is $(R_1 + R_2)/4$.

F 3.5 Insulation testing

Insulation testing is fundamental and will be used as cables are being installed. On completion of the circuit and before energization, the circuit insulation is again checked. The tests show faults or shorts as well as low insulation caused by moisture and similar. Electrical equipment and appliances such as controlgear and lamps should be disconnected prior to testing. Many such devices if left in-circuit would show as an insulation failure; also, sensitive electronic equipment such as dimmer switches and electronic ballasts could be damaged in the test.

Insulation resistance is measured between:

- live conductors, including the neutral;
- live conductors and the protective conductor connected to the earthing arrangement.

It should be noted that Regulation 612.3.1 states that insulation testing is to be made between the live and protective conductors with the protective conductors connected to the earthing arrangement. This is an additional requirement compared with previous editions of the Standard where, for example, cable could be tested to its cpc and then terminated. This procedure may catch you out as you may not be accustomed to carrying it out – so please note.

The basic method of insulation testing is shown in Figures F 3.5 and F 3.6.

Insulation Resistance

Equipment Disconnected

ESSENTIALS

Tests — with CPC connected to each system

Test — between all live conductors to neutral and to earth

Readings — should be approximately tens of MΩ (min 2 MΩ)

Meter

MCB open

Figure F 3.5 Insulation testing.

Insulation - Industrial Application

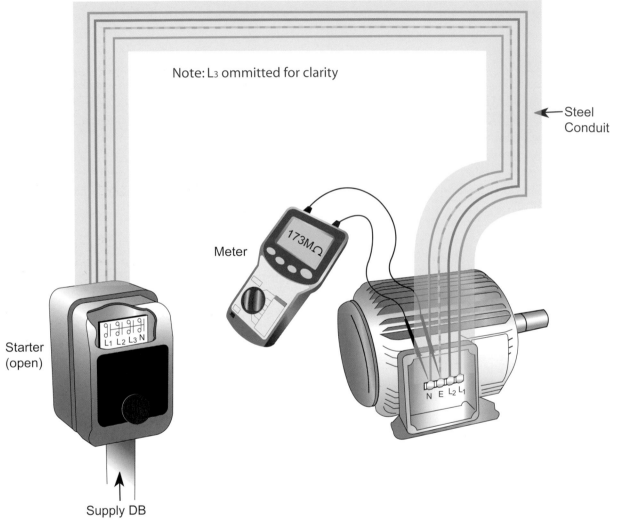

Note: L₃ ommitted for clarity

Steel Conduit

Meter

173MΩ

Starter (open)

Supply DB

Figure F 3.6 Insulation testing of motor circuit.

Table F 3.1 Minimum value of insulation resistance – normal circuits.

Circuit nominal voltage (V)	Test voltage d.c. (V)	Minimum insulation resistance (MΩ)
Normal LV circuits up to 500V	500	≥1.0
Normal LV circuits up to 500V where it is difficult to disconnect sensitive equipment	250	≥1.0

Table F 3.2 Minimum value of insulation resistance SELV, PELV and circuits above 500V.

Circuit nominal voltage (V)	Test voltage d.c. (V)	Minimum insulation resistance (MΩ)
SELV and PELV	250	≥0.5
Above 500V	1000	≥1.0

The minimum values of insulation resistance are given in Table 61 of BS 7671: 2008 reproduced here in Table F 3.1 for normal circuits.

This minimum value of 1 MΩ is an increase from the 0.5 MΩ of the 16th Edition. For most new circuits values would be way in excess of this, usually approaching the maximum scale on the meter.

Lesser used, Table F 3.2 gives minimum insulation resistance values for other circuits.

Perhaps an underused technique is that of carrying out insulation testing on groups of circuits together as shown in Figure F 3.7, and it is recommended that this is limited to 50 outlets per test.

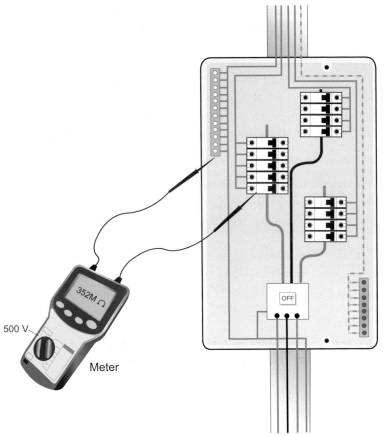

352M Ω

500 V

Meter

ESSENTIALS

Up to 50 outlets in final circuits
can be tested together as
demonstrated here

OFF

Figure F 3.7 Insulation test of a group of circuits.

F 3.6   Polarity testing

Polarity testing is very easy to carry out. There are a few methods and all of the
following are acceptable:

- visual checks where coloured cables are used;
- checks as part of the continuity testing using shorted out cable;
- neon and similar voltage probes;
- multimeters;
- indicators on ELI testers and similar.

Of all of these the visual check is usually the easiest, and this test is one that is best carried out by the person installing and terminating cables, rather than by the 'test engineer'.

Regulation 612.6 requires that every fuse and single-pole control and protective device is connected in the line conductor only. It also requires a check that E14 and E27 lampholders, not to BS EN 60238, have the outer or screwed contacts connected to the neutral conductor; but this does not apply to new installations, as new lampholders should be BS EN 60238 type.

Phase rotation

The correct phasing of three-phase circuits must of course be checked. For all sub-circuits this is usually completed by cable identification and shorting out cables as appropriate. For transformers, generators and the incoming mains LV supply, a phase rotation meter should be used to check for correct phase rotation (Regulation 612.12); follow equipment manufacturers' instructions.

F 3.7   Earth fault loop impedance (ELI) testing

Earth fault loop impedance is required to be checked at various places throughout the installation, and generally at every point where a protective device is installed.

For final circuits, there are two alternative methods of determining the earth fault loop impedance:

1   Direct measurement of total ELI.
2   Measurement of the circuit $R_1 + R_2$ value and addition to the Z_{DB} (earth fault loop impedance at the local distribution board).

Method 2 is not favoured, as explained in Section F 3.3, but if used the general procedure in this section for ELI measurement is followed, and this is added to the circuit $R_1 + R_2$ value, measured as described in Section F 3.3.

F 3.7.1 External earth fault loop impedance (Z_e)

Measurement of external ELI is necessary in LV supplies to confirm the supply earth condition. External ELI is measured live at the intake position, or close to it, with the means of earthing disconnected from the installation and the loop tester connected to it as illustrated in Figure F 3.8. The supply to the installation will need to be isolated. Care should be taken that the installation main equipotential bonding is in place, or that the test is carried out under controlled conditions. The test current on some ELI meters may make exposed-conductive-parts and extraneous-conductive-parts rise in potential in relation to true earth, presenting a

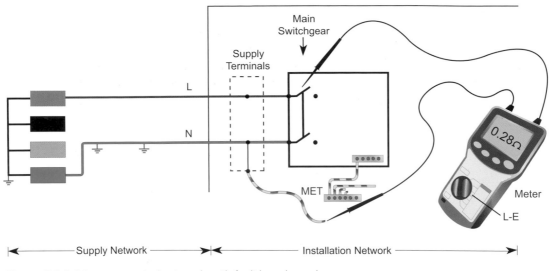

Figure F 3.8 Measurement of external earth fault loop impedance.

potential hazard to persons or livestock. If main equipotential bonding is in place the potential should be no more than a few volts.

Figure F 3.8 is a typical arrangement; in this case for a TN-C-S PME supply. Note that all loop testers illustrated in this book are two lead meters. Three lead loop meters usually perform the same task as two lead units if the neutral and earth leads are connected together, but you should confirm this by consulting the manufacturer's data. Also note that the connection point of the live supply to the meter is not critical to the result.

For UK low-voltage supplies the values should be no more than in Table F 3.3.

Table F 3.3 Maximum external ELI values.

Type of system	External earth fault loop impedance Z_e (Ω)
TN-C-S	0.35
TN-S	0.8
TT	21

Substations

It should be noted that where a transformer is installed on-site, the use of a 'contractors' ELI meter is inappropriate. Small values of resistance cannot be measured on conventional ELI instruments. Confirmation of impedances and d.c. resistance can only accurately be made by calculation and inspection or by the use of 'milli-ohmmeters' utilizing four lead connections. Both methods are outside the scope of this book.

Conventional ELI meters can be very inaccurate at resolutions below 0.1 ohm. Even at readings of 0.2 and similar a digit ± fluctuation has to be applied, and this should be borne in mind when reading the meter at low values (you may have a negative ELI value!).

F 3.7.2 Testing for total earth fault loop impedance (Z_s)

As mentioned in Section C, earth fault loop impedance may be required at various points throughout the installation and will generally need to be measured at every level of protective device.

For confirmation of final circuit disconnection times where RCDs are not installed, measured total earth fault loop impedance is usually required for all circuits.

Z_s may be carried out by direct measurement at the extremity of a circuit. Alternatively, Z_s may be collectively measured using the components in the following formula:

$$Z_s = Z_{DB} + (R_1 + R_2)$$

where:

Z_{DB} is the earth fault loop impedance at the distribution board supplying the final circuit;

$(R_1 + R_2)$ is the circuit measured line-cpc loop resistance (see Section F 3.3).

Whilst carrying out Z_s testing, both the main equipotential bonding and the means of earthing are left connected. This advice was given by the ECA for a number of years in order to maintain consistency between Z_s results when carrying out periodic testing and testing new installations. When carrying out periodic testing, disconnection of earthing and bonding is simply not practicable. The ECA advice was adopted by the IET.

This does mean that in some installations Z_s values measured directly will be less than those measured using the collectively measured $Z_s = Z_{DB} + (R_1 + R_2)$ formula. When comparing the results of measured earth fault loop impedance values with design values, this possible discrepancy should be remembered, as should the matter of possible parallel metallic earth return paths.

The physical measurement of ELI is generally as shown in Figure F 3.9, and it should be remembered that RCDs in circuit will trip unless the test instrument has a facility to block unwanted tripping. Some ELI testers can test at such small current levels that they are below the threshold of tripping. Alternatively, the RCD must be linked-out of the circuit.

The measured values of Z_s should be less than the values given in Chapter 41 of BS 7671: 2008 Tables 41.2, 41.3, 41.4 and 41.5, which are reproduced in Appendix 3. It should be noted that the limiting ELI values given in these tables are design values and should be de-rated by a factor of 0.8.

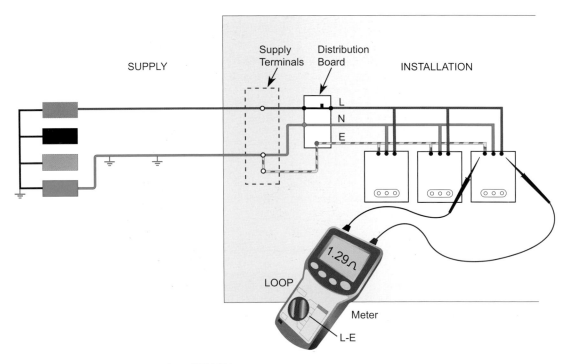

Figure F 3.9 Measurement of total ELI (Z_s).

F 3.8   Prospective fault current testing

Regulation 612.11 requires that the prospective fault current, I_{PF}, under both short-circuit and earth fault conditions, be measured, calculated or determined by another method, at the origin and at other relevant points in the installation. For domestic and LV supplies the utility values can be used and measurement is not required (the utility value is a maximum of 16 kA).

Fault current measurement, however, is quite easy to measure for modest fault levels. Most earth fault loop impedance meters double as fault current measuring devices, capable of measuring phase-neutral prospective fault current as well as earth fault current. Figure F 3.10 shows measurement at two positions within an installation.

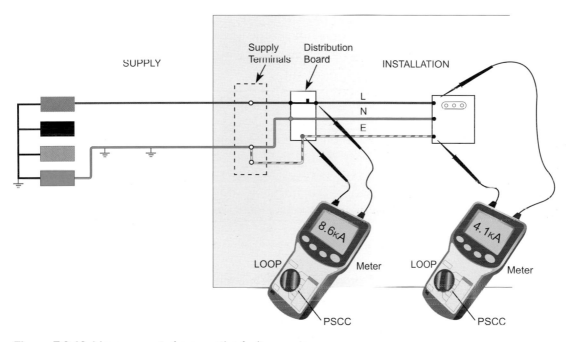

Figure F 3.10 Measurement of prospective fault current.

Bear in mind that these instruments simply measure the loop impedance and, by Ohm's law, calculate the fault current. This will give an overestimate of fault current as the fault will have a low power factor, not taken into consideration by the ELI meter.

Some test instruments are capable of measuring line-to-line fault current. These work by measuring the line-to-line loop impedance. If these instruments are not used the three-phase fault current is approximated as twice the single-phase level.

F 3.9   Testing RCDs and other functional tests

Regulation 612.13.2 requires that switchgear and control gear assemblies, controls and interlocks be functionally tested to check that they work as required. This is part of the commissioning of the electrical installation and not part of the subject of this book.

RCD testing

RCDs should have their functionality tested by operating the integrated 'trip' button. BS 7671: 2008 only requires operating time testing of RCDs where fault protection is provided by an RCD. Where final circuits have overcurrent protection by a fuse or MCB, here are a couple of options:

- either test the RCD and rely on this to achieve the disconnection time (usually within 0.4 s); or
- test the circuit for Z_s and if compliant with the tables in Chapter 41, RCD testing is not essential.

In similar fashion to other daring statements in this book, some may be uncomfortable with this suggestion, but you should consider that RCDs are a manufactured 'type-tested' device; we do not routinely test the operating characteristics of MCBs.

RCD testing itself is relatively simple in terms of using your test instrument connected on the load side of the RCD; Figure F 3.11 shows a typical RCD test.

Interpreting results, especially with time-delayed RCDs, can be a little confusing. Table F 3.4 summarizes the maximum RCD trip times for comparison with measured results and includes minimum trip times for time-delayed RCDs.

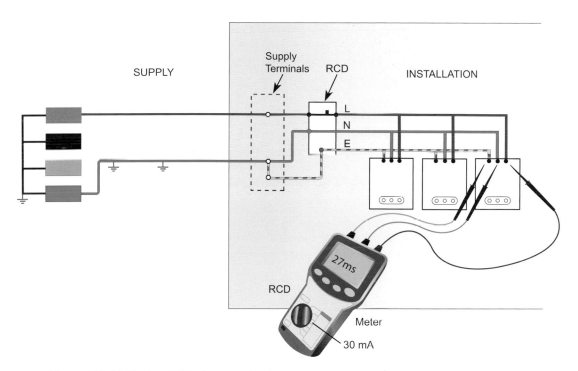

Figure F 3.11 Testing RCD trip operating times.

Table F 3.4 Trip times for RCDs.

RCD type	Sensitivity (mA)	Test current 1 × Δ*n* (mA)	Trip time (ms)	Test current 5 × Δ*n* (mA)	Trip time (ms)
General	10	10	200 max	50	40 max
	30	30		150	
	100	100		500	
	300	300		1500	
	500	500		2500	
Delay type S	100	100	130 min 500 max	500	40 min 150 max
	300	300	130 min 500 max	1500	40 min 150 max
	500	500	130 min 500 max	2500	40 min 150 max

Note: Testing should include a 50% no trip test

F 3.10   Verification of voltage drop

Notes have been included on this subject because it appears in Part 6 of BS 7671: 2008 and gives rise to some confusion.

Regulation 612.14 gives two methods for checking voltage drop, either by measuring a circuit's impedance or by checking design criteria.

The regulation only suggests doing this where it is necessary to verify compliance with the voltage drop requirements. In practice, this will mean where there is a voltage drop problem.

The single most important factor that will affect voltage drop under running conditions is the running current, of both the circuit and the whole installation.

F 4   Certification paperwork

F 4.1   Introduction, various certificates and schedules

Part 6 of BS 7671: 2008 requires various certificates and schedules to be issued on completion of the installation as follows:

- a completion certificate; or
- periodic report form where undertaken; and
- accompanying either of these, results of inspection and test.

This chapter provides in Section F 4.2 a brief overview of the various certificates and accompanying schedules, and a line-by-line brief of how to complete the various forms, in Section F 4.3.

Section F 4.3 describes typical certificates and forms and A4 versions of these are available in Appendix 17. The models suggested are as in BS 7671: 2008 with ECA graphics applied.

F 4.2   Overview of certificates and schedules

Electrical installation certificate (BS 7671: 2008)

This certificate should be used for new installations including alterations and additions. The certificate requires signatures for the three different aspects of design, construction, inspection and test. The certificate should be accompanied by the appropriate schedules of inspections and test results.

Single signatory electrical installation certificate (BS 7671: 2008)

This is a certificate complying fully with BS 7671: 2008 and is used where one individual or organization has been responsible for the all the aspects of design, construction, inspection and testing. The form is combined with a test schedule comprising a maximum of 18 circuits and a schedule of inspections.

Minor electrical installation works certificate (BS 7671: 2008)

For certification of installation work that does not include the introduction of new circuits, e.g. the addition of a socket outlet or lighting point to an existing circuit.

Schedule of test results

This schedule includes all the relevant information from the 'everyday tests' described in Section F 3.

Schedule of inspections

This form is used by the person carrying out the visual inspection and confirms the visual inspections undertaken.

Periodic inspection report for an electrical installation (BS 7671: 2008)

This is discussed in Appendix 15.

F 4.3   Completing the paperwork

This section has been added for completeness and may be skipped by some readers.

F 4.3.1 Electrical installation certificate (BS 7671: 2008)

(For A4 versions, see Appendix 17.)

Details of the client
Insert client's name and title.

Installation address
Insert address of installation.

Description and extent of installation
Describe the extent and limitation of the certificated work.

Tick, as appropriate, the box for either new installation, addition to an existing installation or alteration to an existing installation.

Design, Construction, Inspection & Testing
The appropriate sections should be completed and signed by competent personnel authorized by those responsible for the work of design, construction, inspection and testing respectively. The relevant amendment date of BS 7671 must be added. Any departures from BS 7671 must be indicated.

Next inspection
Add the appropriate recommended date of next inspection – see Section F 4.3.6.

Particulars of signatories
To be completed by the organizations responsible for each aspect of design, construction, inspection and test.

Supply characteristics and earthing arrangements
Earthing arrangement
Add tick to the appropriate box noting the (external) supply characteristic.

Number and type of live conductors
Tick appropriate box(es).

Nature of supply parameters
The nominal supply voltage between phases (U), the voltage to earth (U_o) and the frequency must be added after confirmation by the supply company. The prospective fault current, being the larger of the short-circuit current and the earth fault current, shall be recorded – this is determined either by enquiry, measurement or calculation. The external earth fault loop impedance, Z_e, shall be recorded and is determined either by enquiry, measurement or by calculation.

Supply protective device characteristic
Type – add type of supply protective device. Also add the protective device current rating.

Particulars referred to in the certificate
Method of fault protection
Either tick EEBADOS (Earthed Equipotential Bonding and Automatic Disconnection of Supply) or, if this method is not used, enter 'see attached' in space provided and include the necessary details with the certificate.

Means of earthing
Tick appropriate box – see Chapter E.

Maximum demand
The maximum demand is the designer's estimation of the maximum load demand of the installation expressed in kVA or amps per phase and takes into account diversity. Further guidance on this subject can be found in Chapter C of this book.

F 4.3.2 Single signatory electrical installation certificate (BS 7671: 2008)

(For A4 versions, see Appendix 17.)

Certificate overview
Where design, construction and inspection and testing are the responsibility of one person or organization, this single signatory certificate can be used and fully complies with BS 7671: 2008. It also combines a test sheet and can be used for an installation of up to 18 circuits of any current rating.

Details of client
Insert client's name and title.

Installation address
Insert address of installation.

Description and extent of installation
Briefly describe the installation and describe the extent and limitation of the certificated work.

Tick, as appropriate, the box for either new installation, addition to an existing installation or alteration to an existing installation.

Design, construction, inspection and testing
This section should be completed and signed by the individual responsible for the work of design, construction, inspection and testing of the installation. The relevant BS 7671 amendment date should be added.

Next inspection
Add the appropriate recommended date of next inspection – see Section F 4.3.6.

Supply characteristics and earthing arrangements
Voltage
Add supply voltage to earth (U_o) and supply frequency in Hz. The prospective fault current must be recorded, and is the larger of the short-circuit current and earth fault current established by enquiry or measurement. The external earth fault loop impedance Z_e shall be recorded and may be measured or determined by enquiry.

Number and type of live conductors
Tick appropriate box(es).

Supply protection
Type – add type of supply protective device. Also add the protective device current rating.

Earthing arrangement
Tick appropriate box noting the (external) supply characteristic.

Supply earth and earth electrode
Where there is an electricity company earth, tick the box. If there is a private earth electrode system, provide details here.

Particulars of installation referred to in this certificate
Method of fault protection
The certificate can only be used for normal ADOS (Automatic Disconnection of Supply) and this method has been pre-ticked to indicate this.

Maximum demand
The maximum demand is the designer's estimation of the maximum load demand of the installation expressed in kVA or amps per phase and takes into account diversity. Further guidance on this subject can be found in Chapter C.

Main switch or circuit breaker
Complete all entries.

Main protective conductors
Complete as necessary.

Submain
Add details of submain where applicable.

Distribution board
Add details – Z_s and prospective fault current are measured (or calculated) at the distribution board. Note: do not add external readings.

Installed circuit details
A   Add circuit reference.
B   Describe circuit briefly, i.e. ring, socket outlets.

Top of columns C & D
Add short circuit breaking capacity of overcurrent device as follows:

BS	Type	Typical breaking capacity (kA)
BS 88	HRC cartridge fuse	80
BS 3036	Rewireable	2
BS 1361	Household cartridge fuse	16.5
BS EN 60898	Circuit breaker	3, 6, 9, or 16

C   Add type of protection device. For fuse add BS number, for circuit breaker add sensitivity type B, C or D.
D   Add current rating of protective device.
E   *Optional column.*
         Add installed reference method from Appendix 4 of BS 7671.

F Add line and neutral conductor size where they are the same. If reduced neutral add to remarks column.

G Add cpc size. Optionally, for armoured cables, add cross-sectional area of armour in mm². If not SWA add material type.

H Continuity. Add maximum value obtained from method used to check continuity of cpc at all points on circuit and delete $R_1 + R_2$ or R_2 as appropriate, depending upon method used.

Optionally, this column may be ticked.

Where the circuit is a ring the $(R_1 + R_2)$ value must be inserted into column H and will be the $(R_1 + R_2)$ on the ring with the phase and cpc cross-connected at the board. See Section F 3.4.

J, K & L ring continuity only

J Add open phase/phase resistance.

K Add open neutral/neutral resistance.

L Add open cpc/cpc resistance.

Insulation resistance

M Test between phase conductors and phase to neutral and record the minimum.

N Test phase to earth and neutral to earth either together or separately and record the minimum.

P *Polarity.* Tick when polarity at all points has been checked.

R *Earth fault loop impedance.* Either add Z_e at incomer (distribution board) to the $(R_1 + R_2)$ value or measure at remote part of circuit. Record the maximum value measured.

Functional tests

S Check RCD trip time at normal rate current setting only.

T Tick this column after functional checks are made including: assemblies, switchgear/control gear, drives, controls and interlocks to show they are properly mounted, adjusted and installed in accordance with BS 7671: 2008.

Other comments

Add relevant comments.

Test instruments used

Add details of all instruments used whilst testing.

F 4.3.3 Minor electrical installation works certificate (BS 7671: 2008)

(For A4 version, see Appendix 17.)

This certificate is fully compliant with BS 7671: 2008 when used for the certification of electrical work which does not include a new circuit (e.g. circuit extensions). It should not be used for a consumer unit change.

Description of minor works
Complete as necessary.

Installation details
1   System earthing arrangements – tick appropriate box.
2   Tick boxes or describe other methods as appropriate.
3   Add BS and type of protective device and list its rating.

Add any relevant comments.

Essential tests
Complete these minimum essential tests including the installation phase to neutral test only where practical.

Instruments used
List all instruments used while testing.

Declaration
Sign as necessary.

F 4.3.4 Schedule of test results

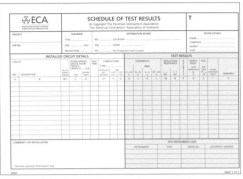

(For A4 version, see Appendix 17.)

Project
Add details.

Job no.
Optional, for your company use.

Submain
Add details where applicable.

Distribution board
Add details – Z_s and prospective fault current are measured (or calculated) at the distribution board.

Installed circuit details
A   Add circuit reference.
B   Describe circuit briefly, i.e. ring, socket outlets.

Top of columns C & D
Add short circuit breaking capacity of overcurrent device as follows:

BS	Type	Typical breaking capacity (kA)
BS 88	HRC cartridge fuse	80
BS 3036	Rewireable	2
BS 1361	Houshold cartridge fuse	16.5
BS EN 60898	Circuit breaker	3, 6, 9, or 16

C   Add type of protection device. For fuse add BS Number, for circuit breaker add sensitivity type B, C or D.
D   Add current rating of protective device.

E *Optional column*

Add installed reference method from Appendix 4 of BS 7671.

F Add line and neutral conductor size where they are the same. If reduced neutral add to remarks column.

G Add cpc size. Optionally, for armoured cables, add cross-sectional area of armour in mm^2. If not SWA add material type.

H Continuity. Add maximum value obtained from method used to check continuity of cpc at all points on circuit and delete $(R_1 + R_2)$ or R_2, as appropriate, depending upon method used.

Optionally, this column may be ticked.

Where the circuit is a ring the $(R_1 + R_2)$ value must be inserted into column H and will be the $(R_1 + R_2)$ on the ring with the phase and cpc cross-connected at the board. See Section F 3.4.

Ring continuity only

J Add open phase/phase resistance.

K Add open neutral/neutral resistance.

L Add open cpc/cpc resistance.

Insulation resistance

M Test between phase conductors and phase to neutral and record the minimum.

N Test phase to earth and neutral to earth either together or separately and record the minimum.

P *Polarity.* Tick when polarity at all points has been checked.

R *Earth fault loop impedance.* Either add Z_e at incomer (distribution board) to the $(R_1 + R_2)$ value or measure at remote part of circuit. Record the maximum value measured.

Functional tests

S Check RCD trip time at normal rate current setting only.

T Tick this column after functional checks are made including: assemblies, switchgear/control gear, drives, controls and interlocks to show they are properly mounted, adjusted and installed in accordance with BS 7671: 2008.

V Add any relevant remarks to each circuit.

Other comments

Add relevant comments to section.

Test instruments used

Add details of all instruments used whilst testing.

F 4.3.5 Schedule of inspections

(For A4 version, see Appendix 17.)

Simply insert a tick into the appropriate box indicating the inspections made. Add a 'N/A' to the box where the inspection is not applicable.

You may feel it is appropriate to insert 'LIM' (for limitation) into a box where the inspection type is limited to certain areas. If this is the case, you should create your own 'LIM' legend on the schedule.

The schedule should be used with an associated Electrical Installation Certificate or Periodic Inspection Report.

F 4.3.6 Recommended frequencies of next inspection

The frequencies of 'next' inspection are shown in Table F 4.1 and reproduced from IEE Guidance Note 3.

Table F 4.1 Recommended initial frequencies of electrical installation inspections.

Type of installation	Maximum period between inspections and testing as necessary	Reference (see notes below)
Domestic	Change of occupancy/10 years	
All commercial i.e. shops, offices, hospitals and labs etc.	Change of occupancy/5 years	1
Industrial	3 years	
Places subject to entertainment licence	1 (note 2)	1,2
Public swimming pools, caravan parks	1 year	1,2

[1] See also Electricity at Work Regulations 1989.
[2] This is normally a requirement of local licensing organizations.

Special Locations

G 1   Introduction: Purpose and principles

G 1.1   Introduction

Part 7 of BS 7671: 2008 makes specific references to electrical installations in areas defined as 'Special Installations or Locations'. As well as having to comply with the general requirements of BS 7671, some further requirements or restrictions apply to installations in these special locations. The term 'special' therefore means application of particular requirements in addition to the general rules of BS 7671 within Parts 1 to 6.

BS 7671 introduces some six new 'Part 7' special locations making a total of 14 in all, and the complete list is shown in Figure G 1.1. The provision of special locations sections has increased in IEC over the years, and it may be argued that in some cases they are not really 'special', but they have been accepted internationally.

Some of the special locations in BS 7671: 2008 locations are not discussed in this book, as most designers and contractors do not get involved with them. Figure G 1.1 shows the Part 7s discussed in this book and those that are not.

Some requirements have been removed from those provided in the 16th Edition, for example, Section 607 'High protective conductor currents' and Section 611 'Installation of highways power supplies, street furniture and street located equipment'; these sections have both been incorporated into the general rules, i.e. 543.7 and 559.10, respectively, of the 17th Edition.

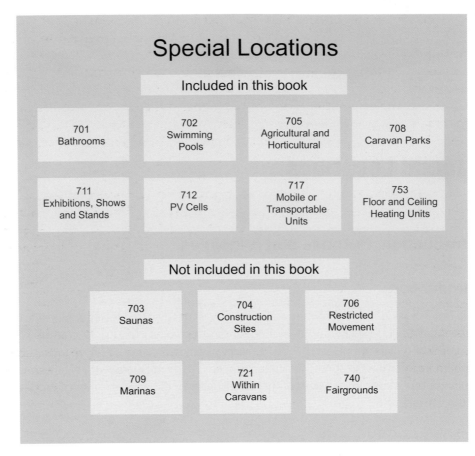

Figure G 1.1 Part 7 sections showing those discussed in this book.

G 1.2 Purpose and principles

In general, such 'Special Installations or Locations' listed in BS 7671 involve increased risks and/or harsher conditions compared with those catered for by the general parts (Parts 1–6). Risk here is risk of *danger* as defined in part 2 of the Regulations:

● Onerous site conditions that may impair the effectiveness of the electrical or environmental protective measures.
● Increased risk of contact with earthed metalwork, usually extraneous-conductive-parts.
● Water or condensation giving reduced body resistance and better electrical contact.
● Absence of shoes or clothing (e.g. bathing or swimming) and therefore a greater opportunity to make contact with live parts or earthed metalwork, or actual contact with Earth itself.

The requirements for such locations have been developed by continual assessment of such risks, and the application of protective and environmental measures seen to be necessary to reduce the foreseen risks. Designers and installers must assess the environment, use and risks at the locations for which they are designing and constructing installations, and make provisions accordingly – a high level of competence and relevant experience is required. Designers and installers must also comply with the requirements of legislation and relevant building regulations. Installations must be properly verified and commissioned, with initial inspection and testing as required by Part 6 of BS 7671.

All installations must be properly maintained and periodically inspected and tested, depending on their age, application and condition. An assessment of the expected maintenance that an installation will receive should be made as part of the design process, and if necessary the design and materials selected should be amended accordingly.

G 1.3 Particular requirements and numbering

It is the case that Part 7s are intended to supplement or modify the general requirements of Parts 1–6. In line with this is new numbering, which is meant to highlight the particular requirements; this uses the Part 7 numbering followed by the 'general rules' clause. An example from 701 (bathrooms) is as follows:

701.41 Protection for safety: protection against electric shock

701.410.1 General requirements

701.410.3.5 The protective measures of obstacles and placing out of reach (Section 417) are not permitted.

Regulation 410.3.5 in Part 4 of the Standard has been modified for this Part 7 Regulation 701.410.3.5 above.

G 2 Locations containing a bath or shower (701)

G 2.1 Introduction and risks

Bathroom installation regulations have perhaps always been somewhat contentious and have undergone a number of changes in requirements in the last ten years. The requirements have again changed in the 17th Edition of BS 7671 and indeed have become simpler. This part now harmonizes with the corresponding CENELEC part, and introduces some significant changes for the UK.

In a bathroom or shower room the increased risks are from exposed wet skin (with a lower contact resistance), the splashing and ingress of water, and the usual close proximity of earthed metalwork. There may also be further risks due to disability or infirmity. Medical locations are not considered in detail, and such medical locations as birthing pools need special provisions outside the scope of this document.

G 2.2   Zone concept

The concept of zones was introduced into BS 7671 for the 2001 edition and allowed different rules to be applied to different areas. The concept of zones remains in BS 7671: 2008 but Zone 3 has been removed. This is one of the most significant regulation changes in the whole document; without a Zone 3, there are no specific rules for Zone 3 and only the general rules apply. Hence, equipment can be installed at the boundary of Zone 2.

The zones provide a means of controlling the location type and environmental rating of electrical equipment installed in a bathroom. They are only a guide and, regardless of this, the installation must be properly designed to take into account all foreseen risks in the design of a bathroom within a particular building. The Regulations actually apply to 'locations containing a fixed bath or shower', and this can therefore include bedrooms, changing facilities in workshops or sports clubs, etc. These will require different design solutions from those for a domestic bathroom, and in all cases full consultation with the client's advisers and equipment manufacturers is essential. Unfortunately, not all electrical equipment or accessories have IP ratings, but manufacturers will be able to provide guidance in the selection and application of their products.

The zones are specified in regulations and are summarized as follows:

● Zone 0 is the bath or shower tray.
● Zone 1 can be considered to be the area where the individual is bathing or showering, or the area where shower water is likely to be directly sprayed.
● Zone 2 is the area beyond Zone 1, extending by a further 600 mm.

The layout of the zones is illustrated in Figures G 2.1 and G 2.2 for a number of bathroom and shower room arrangements. Other layouts such as bedrooms with shower cubicles, changing rooms and the like must be considered; the concepts of the zones should be extended accordingly.

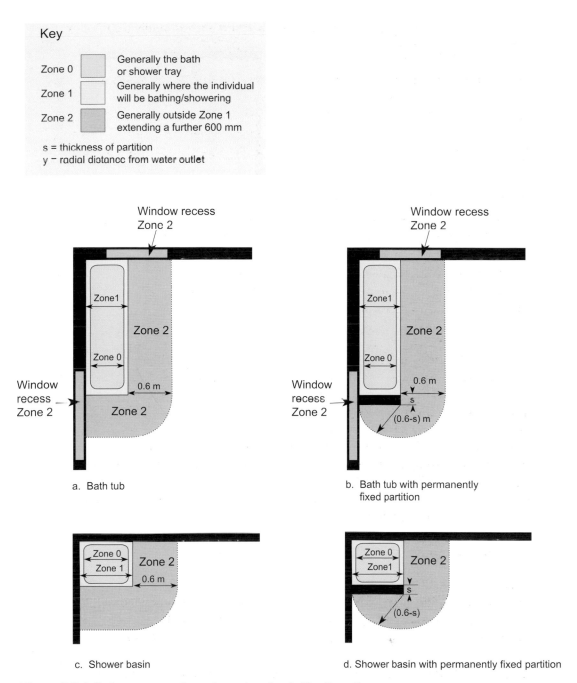

Figure G 2.1 Bathroom zone dimensions plans (a–d). (*Continued.*)

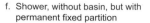

e. Shower, without basin

f. Shower, without basin, but with permanent fixed partition

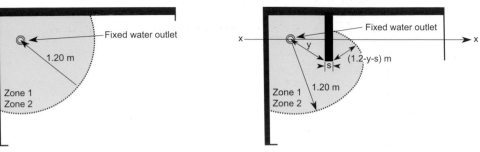

Figure G 2.1 (*Continued.*) Bathroom zone dimensions plans (e–f).

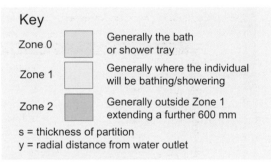

Key

Zone 0		Generally the bath or shower tray
Zone 1		Generally where the individual will be bathing/showering
Zone 2		Generally outside Zone 1 extending a further 600 mm

s = thickness of partition
y = radial distance from water outlet

a. Bath tub

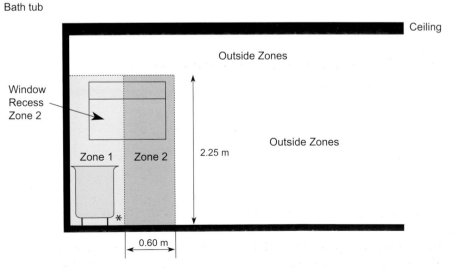

Figure G 2.2 Bathroom zone dimensions elevations (a). (*Continued.*)

c. Shower basin

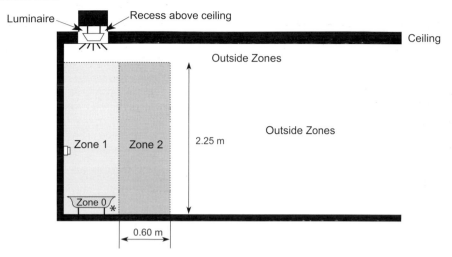

f. Shower without basin, but with permanent fixed partition

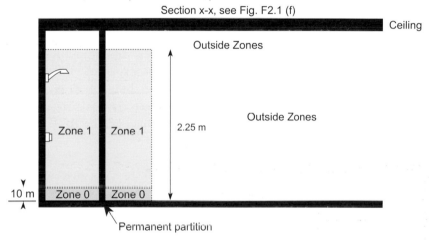

* Zone 1 if the space is accessible without the use of a tool.
Spaces under the bath accessible only with the use of a tool are outside the zones.

Figure G 2.2 (*Continued.*)

For electrical provisions the general requirements of BS 7671 apply, but the extra provisions in Section 701 must supplement these. Obviously, not all the protective measures for installations are applicable in bathrooms; the protective measures for obstacles, placing out of reach, non-conductive locations and earth-free equipotential bonding are not permitted, and it is not difficult to see why.

Table G 2.1 Bathroom electric shock requirements.

Requirement	Regulation
All circuits to have 30 mA RCD	701.411.3.3
Socket outlets to be 3 m from the outer limit of Zone 1 (except SELV)	701.512.3
Where main equipotential bonding is used in the installation, no local supplementary bonding is required	701.415.2
Obstacles and Out Of Reach are not permitted	701.410.3.5
Non-conducting location and earth-free local equipotential bonding are not permitted	701.410.3.6
Electrical separation can only be used for single items	701.413
For SELV and PELV basic protection must be used	701.414.4.5

G 2.3 Electric shock requirements

The requirements in addition to the general rules are summarized in Table G 2.1.

Perhaps the most significant change in this section in the 17th Edition is the fact that there are only regulations for Zones 1 and 2. Outside of Zone 2 the general rules of the Standard apply, but in reality virtually any equipment can now be installed outside Zone 2. The only caveat concerns socket outlets, where Regulation 701.512.3 prohibits them within 3m of the outer limit of Zone 1 (unless they are SELV). Some individuals will feel uncomfortable installing socket outlets even under these conditions (and despite the fact that they will be RCD protected, see below).

All circuits supplying the bathroom are now to be provided with additional protection by the use of an RCD with a rated residual operating current not exceeding 30 mA. However, where this is provided, supplementary equipotential bonding is not required.

The use of RCDs on all circuits, including lighting circuits, may cause some problems initially (e.g. with possible increased risk of hazards if lights go out unexpectedly). These can be overcome relatively easily with some planning; e.g. the bathroom on a separate lighting circuit or two lighting circuits for the bathroom and surrounding rooms.

Protection by electrical separation may be used, but this is a difficult concept to achieve in practice and is usually only provided as a shaver socket outlet; however, the supply to such an outlet will still require RCD protection.

Separated extra low voltage (SELV) systems may be installed, but again these are difficult to achieve in practice, and need to be maintained and controlled by a competent person. Protected extra low voltage (PELV) systems are more common but again difficult to achieve in practice and must be properly designed and specified, so such systems are rarely utilized.

The requirements of the 17th Edition of BS 7671 are not retrospective. This edition, however, now allows that supplementary equipotential bonding is not required in bathrooms provided that each circuit in the bathroom is provided with additional protection by an RCD sensitivity not exceeding 30 mA. This will be the preferred solution in new developments, but there are many existing dwellings and other locations with bath and shower facilities that already have supplementary equipotential bonding, and when bathrooms are modified or refurbished it may be more economic to extend or modify this rather than rewire and install RCDs.

G 2.4   Equipment selection and erection

The equipment requirements of Section 701 are summarized in Table G 2.2, and the requirements fall into two types of regulation – ingress protection and suitability as regards switches and accessories.

Table G 2.2 Bathroom equipment selection and erection requirements.

Requirement	Regulation	Notes
Equipment in Zone 0 shall be IP X7	701.512.2	
Equipment in Zones 1 and 2 shall be IP X4	701.512.2	
Equipment exposed to cleaning jets to be at least IP X5	701.512.2	Particular requirements exist for IP and water jets; see D 16.2
In Zone 0, no switchgear or accessories allowed	701.512.3	
In Zone 0, only 12 V current-using equipment complying with a relevant Standard is allowed	701.55	
In Zone 1, only 12 V SELV switchgear or accessories allowed	701.512.3	
In Zone 1, only whirlpool units, electric showers, shower pumps, ventilation equipment, towel rails, water heaters, luminaires and 25 V SELV or PELV equipment is allowed	701.55	
In Zone 2, SELV switches and socket outlets and shaver supply socket outlets to BS EN 61558–2-5 are the only allowed switchgear or accessories	701.512.3	

The general principle of this group of regulations is that unsuitable electrical equipment must be inaccessible to persons in the bath or shower (when they are in the bath or shower, Zone 1). This is deemed to be outside Zone 2.

This 'inaccessibility' principle extends to electrical equipment generally, and only switches using insulated linkages or pull cords to operate BS 3676 devices, or specially designed controls to BS 3456 (instantaneous water heaters), are permitted within the zones. It is accepted, however, that a shaver supply unit with a BS EN 61558–2–5 transformer may be installed, as may switches supplied by SELV where the nominal voltage does not exceed 12 V rms.

Fixed equipment within the 2.5 m zone must be selected and erected according to foreseen risks, and its likely duty. The likelihood is for splashing water to be present, and this requires an IP rating dependent on its location in the room. Power showers can generate penetrating jets of water, and may require a minimum of IPX5 equipment.

BS 7671 does not actually prohibit the use of Class I fixed equipment for outside of Zone 2, but luminaires to be fixed within these areas should preferably be either totally enclosed, or if of Type B22, should be fitted with a BS 5042 (Home Office) shield.

In general, those responsible for selection and erection of equipment will find that Class II devices more readily satisfy the requirements of bathrooms.

With the increased use of pumped water in bathrooms for spa-baths, power showers, etc. there is a need, on occasion, to provide for motive power within the area of the bath. In order to protect against direct and indirect contact, supplies for such equipment must be by SELV with the nominal voltage not exceeding 12 V rms. Where it is necessary to mount the SELV source within the bath enclosure, then it may only be accessible by means of a tool. The reason for this is concerned with the now familiar definition of a skilled person, i.e. it is assumed that persons using a tool to access the SELV source under a bath will be sufficiently informed and skilled to avoid danger and that the circuit will be dead and isolated before work is commenced.

G 3   Swimming pools and other basins (702)

G 3.1   Introduction and risks

BS 7671: 2008 maintains the next Special Location in Part 7 as 'Swimming Pools and Other Basins'. The 'other basins' is important as the section applies to the basins of fountains and to areas in natural waters including the sea and lakes, where they are specifically designated as swimming areas.

Swimming pools and basins pose similar risks to bathrooms, in that people are generally unclothed and body contact resistance is low. One major difference to keep in mind, however, is that generally there are several people in the pool area and usually only one person in the bathroom.

Some fountain basins are likely or even expected, to be occupied, and as such they should be treated as swimming pools as far as BS 7671 is concerned. Often a risk assessment will need to be carried out to assess whether fountains need to be treated as fountains or as swimming pools.

G 3.2   Zone concept

Again, the swimming pool or fountain basin is first divided into zones in order that regulatory requirements can be prescribed for each zone. The zones for swimming pools are shown in Figures G 3.1 and G 3.2 (pools above and below ground).

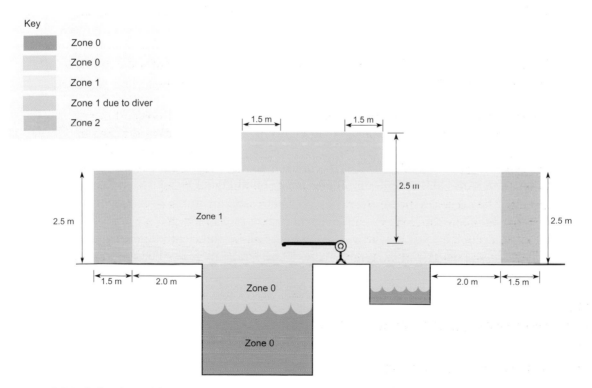

Figure G 3.1 Swimming pool zones.

Key

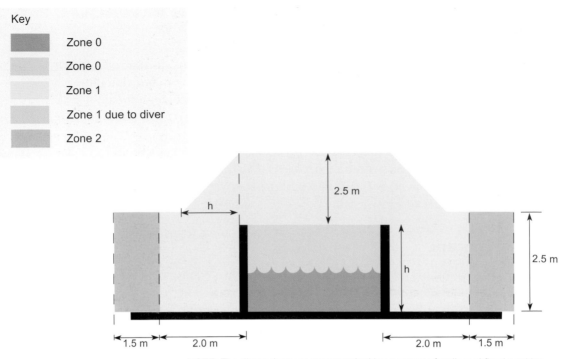

NOTE: The dimensions are measured taking account of walls and fixed partitions

Figure G 3.2 Swimming pool zones – raised pool.

Also, for fixed partitions the zone plan in Figure G 3.3 will be useful.

For fountains, there is no Zone 2 as it is unlikely that persons will be unclothed in public fountains. The zones for fountains are given in Figure G 3.4.

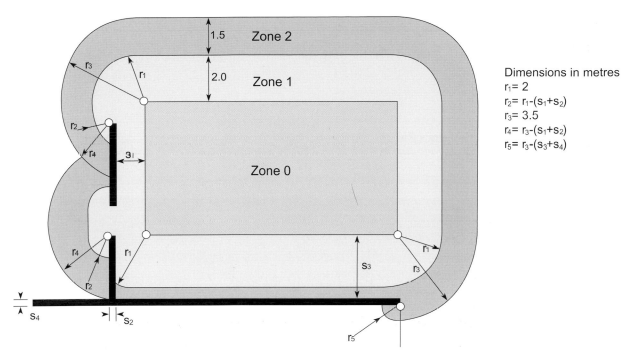

Dimensions in metres
$r_1 = 2$
$r_2 = r_1 - (s_1 + s_2)$
$r_3 = 3.5$
$r_4 = r_3 - (s_1 + s_2)$
$r_5 = r_3 - (s_3 + s_4)$

Figure G 3.3 Swimming pool zones – from various fixed partitions.

Key

Zone 0

Zone 1

Zone 0 Air spray below waterjets and waterfalls to be considered as zone 0

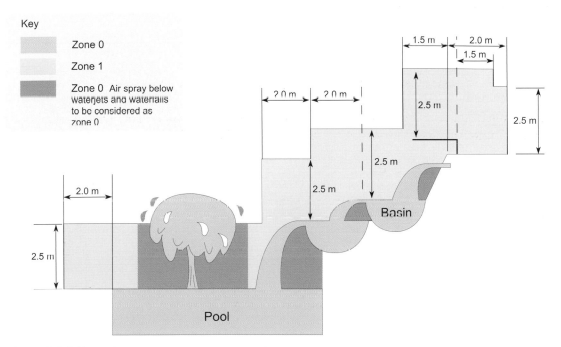

Figure G 3.4 Fountain zones.

G 3.3 Requirements and guidance

The requirements additional to the general rules are summarized in Tables G 3.1 and G 3.2.

Table G 3.1 Swimming pool and fountain electric shock requirements.

Swimming pool or fountain	Requirement	Regulation
Both	All extraneous-conductive-parts in Zones 0, 1 and 2 (except for fountains) to be supplementary bonded	702.411.3.3
Both	For SELV and PELV basic protection must be used	702.414.4.5
Swimming pools	Only SELV up to 12 V a.c. or 30 V d.c. can be used as a protective measure in Zone 0 in swimming pools	702.410.3.4.1
Swimming pools	Only SELV up to 25 V a.c. or 60 V d.c. can be used as a protective measure in Zone 1 in swimming pools	702.410.3.4.1
Swimming pools	Protective measure in Zone 2 in swimming pool can be SELV, RCD automatic disconnection or 'single item electrical separation'	702.410.3.4.3
Fountains	Protective measure in Zones 0 and 1 in fountains can be SELV, RCD automatic disconnection or 'single item' electrical separation	702.410.3.4.2

Table G 3.2 Swimming pool and fountain equipment selection and erection requirements.

Swimming pool or fountain	Requirement	Regulation	Notes
Both	Equipment in Zone 0 shall be IP X8	702.512.2	
Both	Equipment in Zone 1 shall be IP X4	702.512.2	
Both	Equipment exposed to cleaning jets to be at least IP X5	702.512.2	Particular requirements exist for IP and water jets, see D 16.2
Both	Surface metallic wiring sheaths or conduits, or those less than 50 mm deep shall be connected to the local supplementary bonding	702.522.21	
Both	In Zones 0 and 1, no switchgear is permitted	702.53	
Both	In Zones 0 and 1, no socket outlet is permitted	702.53	
Swimming pool	In Zones 0 and 1, only equipment designed for a swimming pool can be used	702.55.1	
Swimming pool	In Zones 0 and 1, floor heating unit is allowed if it is SELV, RCD protected and covered with earthed metal grid	702.55.1	
Both	Luminaires shall comply with BS EN 60598–2-18	702.55.2 702.55.3	

Table G 3.2 (*Continued.*)

Swimming pool or fountain	Requirement	Regulation	Notes
Swimming pools	Zone 1 pool equipment (pumps etc.) shall be in a Class II enclosure, only accessible with a tool and have 25 V a.c SELV (60 V d.c.) or electrical separation	702.55.4	
Swimming pools	For swimming pools without a Zone 2, luminaires can be installed in Zone 1 and do not need to be 12 V SELV, but must be RCD protected and at least 2 m above floor	702.55.4	

To further assist with the tables, the following notes have also been included for extra clarification and guidance.

Apart from underfloor heating there is no specific requirement for a metallic grid in the floor (this was deleted by amendment of the 16th Edition). Connections to the protective conductors of exposed-conductive-parts may be made locally to the equipment or at a local distribution board or control panel, depending on the installation design.

Extraneous-conductive-parts are conductive parts not forming part of the electrical installation and liable to introduce an electric potential, including the electric potential of a local earth. For Section 702 this means a potential from outside Zones 0, 1 and 2 into these Zones. Such parts may include the following:

● metallic pipelines for fresh water, waste water, gas, heating, climate and other;
● metallic parts of building construction;
● metallic parts of the basin construction;
● reinforcement of non-insulating floors;
● reinforcement of concrete basins.

Floors made of individual concrete tiles whose reinforcement is fully encapsulated within the tile and not accessible without damaging the tile, need not be included in supplementary equipotential bonding.

Concrete tiles without metallic reinforcement, tile coverings as well as topsoil (e.g. lawn) need not be included in supplementary equipotential bonding.

The following conductive parts are not regarded as extraneous-conductive-parts provided that they cannot introduce a potential to Zones 0, 1 and 2; they need not be included in supplementary equipotential bonding:

- basin ladders;
- basin barriers;
- diving structures, ladders;
- handrails and handholds on the rim of the basin;
- grid covers including the mounting frames of overflow pipes;
- window frames;
- door frames;
- starting blocks;
- other similar items.

For fountain basins, cables run to supply lighting equipment in Zone 0 are to be run outside the basin as much as possible, and take the shortest practicable route in Zone 0 to the equipment. Cables must be suitable for total continuous immersion and type H07RN8-F is recommended. If it is possible for people to get into the basin, cables should not be installed.

Socket outlets are not permitted in Zones 0 or 1 of a swimming pool area, and are normally only permissible in Zone 2 if they are supplied either by SELV, the source being installed outside the zones, or by the application of electrical separation; again the transformer being outside the zones, or protected by a 30 mA RCD.

Pool cleaning equipment at mains voltage or special equipment should only be brought into the pool area when it is empty of swimmers, and supplied from sockets outside the zones.

G 4 Agricultural and horticultural premises (705)

Adrian Peacock/Getty Images/Digital Vision

G 4.1   Introduction, purpose and principles

BS 7671: 2008 includes this Section 7 for fixed electrical installations inside and outside agricultural and horticultural buildings, including locations where animals are kept.

It does not apply to residences or other locations such as shops, workshops or storage areas in agricultural premises.

These locations are characterized by arduous conditions, and people usually working in a wet or damp environment, both indoors and outdoors. An increased risk of damage to the electrical installation and an increased risk of personal danger can come from a number of factors, including:

- the use, possibly widespread, of chemical cleaners and fertilizers;
- behaviour and nature of animals (stock and vermin);
- agricultural machinery;
- frequent wet and damp conditions;
- lower body resistance of livestock.

It should be noted that danger, including the risk from electric shock, particularly applies to livestock. In places where livestock is kept, there is a greater risk of electric shock to them due to their lower body resistance and more intimate contact with the general mass of earth. It can often be seen in milking parlours that cows will not pass from one place to another where they sense a small potential difference between their front and rear legs.

G 4.2   Requirements and guidance

The requirements of the section are summarized in Table G 4.1.

As well as Table G 4.1 the following notes and guidance are provided.

As applies generally to special location areas, protection by the use of obstacles, placing out of reach, non-conducting locations or earth-free equipotential bonding is not allowed in agriculture/horticulture due to the extra risks involved in such wet and exposed locations, as operators are usually non-technical staff (ordinary persons).

Wiring systems and electrical equipment and accessories must be suitable for the location and environment of their use. Arduous conditions including animal housing, cleaning chemicals, wash down with hoses, vermin, physical impact damage

Table G 4.1 Agricultural and horticultural installation requirements.

Requirement	Regulation
Final socket-outlet circuits with socket outlets up to 32 A require a 30 mA RCD	705.411.1
Final socket-outlet circuits with socket outlets above 32 A require a 100 mA RCD	705.411.1
All other circuits require an RCD not exceeding 300 mA	705.411.1
For SELV and PELV basic protection must be used	705.414.4
Where livestock is kept, supplementary bonding shall connect exposed and extraneous-conductive-parts; this includes floor reinforcing Copper conductors to be 4 mm² minimum, steel bonding to be galvanized and minimum of 8 mm diameter or 30 x 3 mm section	705.415.2.1 705.544.2
Where welfare of animals is affected by loss of supply (e.g. food, water, ventilation or lighting systems) a standby supply shall be installed and separate final circuits shall be used. Alternatively for ventilation systems, monitoring and alarms can be used	705.560.6
Ventilation supply circuits shall be designed to achieve discrimination	705.560.6
Electrical heaters for livestock shall be to BS EN 60335–2-71	705.422.6
All equipment to have IP minimum of IP44	705.512.2
Obstacles and Out Of Reach are not permitted	705.410.3.5
Non-conducting location and earth-free local equipotential bonding are not permitted	705.410.3.6

etc. are all possible, and the selection and erection of systems and equipment must provide proper protection and safety.

All electrical equipment should have physical protection to withstand wash down and the use of chemical cleaning agents, and will require special considerations against corrosion. High impact plastic materials may be better than metals, but these cannot be considered to be a Class II installation. Generally, the best physical protection will be provided by careful placement of outlets and controls in areas where they are not likely to be subject to damage or impact.

Electrical equipment must be protected against the ingress of both solid particles and water, depending upon its location, and BS 7671 recommends a minimum rating of IP44 under normal conditions; equipment not so rated should be enclosed within an enclosure rated to at least IP44. This may, however, not be practical for all items, such as sensors or controls, and other alternatives, such as locating these in safer areas (if possible) may be necessary. IP44 is really a general minimum, and an environmental assessment must be made at the design stage and equipment selected accordingly. It must be noted that IP ratings are not 'all inclusive' and their application to various environments must be considered. For example, IPX8 may be suitable for total immersion, but it may not be suitable for water jets and areas being hosed down (see D 16.2).

In areas with no specific risk, e.g. residential areas, offices, shops and similar locations belonging to agricultural and horticultural locations, a normal residential level of protection should be adequate, but if there is any extra risk, e.g. the practice of using extension leads into other areas (which should be avoided), further provisions may be necessary.

Special consideration must be given to earthing and bonding to ensure its integrity, and stranded conductors are recommended, especially in areas where vibration is likely to be experienced.

Earthing and protective bonding conductors must also be protected against damage and corrosion, and BS 7671 suggests minimum cross-sectional areas of conductors of various materials (see Chapter E).

The reduced disconnection time of 0.2 s of the 16th Edition has been deleted for the 17th Edition, and there is now no difference between the special locations disconnection time requirements and those specified in Chapter 41. Again, the use of RCDs is seen as a positive safety benefit, but the designer must consider the possible consequences of loss of electrical supply to farm buildings or systems that may provide life support for animals or other safety services. Indeed, the requirements for agricultural and horticultural installations in the 17th Edition call for the consideration of supply security and possible standby supplies for animal life-support or comfort systems.

The general requirements of Part 4 of BS 7671 apply, and in all circuits, whatever the type of earthing system, 30 mA RCDs (to the requirements of Regulation 415.1.1) are to be provided for all socket-outlet circuits up to and including 32 A socket outlets, and a 100 mA RCD (again, to the performance requirements of Regulation 415.1.1) must be provided for circuits with socket outlets over 32 A.

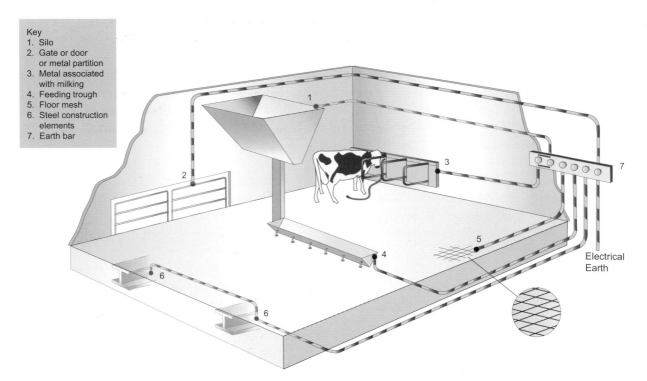

Key
1. Silo
2. Gate or door or metal partition
3. Metal associated with milking
4. Feeding trough
5. Floor mesh
6. Steel construction elements
7. Earth bar

Electrical Earth

Figure G 4.1 Required earth bonding in milking parlour and similar.

In livestock locations all exposed-conductive-parts and extraneous-conductive-parts that are accessible to livestock must be connected together with supplementary equipotential bonding. Figure G 4.1 gives an illustration of the requirement.

Where a metal grid is installed in the floor, or there are extraneous-conductive-parts in or on the floor, e.g. structural steel components, or cattle pens with 'cast in' items embedded in the concrete, or concrete reinforcing, they must be connected to the supplementary equipotential bonding. The illustrations in Figure G 4.1 are only for general guidance and other suitable bonding arrangements are quite acceptable.

If a bonded metal grid cannot be laid in the floor of animal houses during construction or refurbishment, it is recommended that a TN-C-S (PME) electricity supply is not used. A TN-S supply is preferred. If a TN-C-S supply is offered or is the only type available, it is recommended that a separate local earth electrode be installed.

Fire has always been a significant risk in agricultural and horticultural buildings, with the amount of flammable materials used and stored in them. A lot of fires are put down to 'electrical faults' and, although the use of RCDs cannot be guaranteed to prevent the risk of electrically ignited fires, correctly selected RCD protection can reduce electrical fires caused by earth leakage current ignition. For fire protection purposes, an RCD with a rated residual operating current not exceeding 300 mA is advised, and a lower operating current may be appropriate depending on the installation and loads supplied.

Lastly, if electric fences are installed, details of operation and maintenance should be provided to the user of the installation.

G 5   Caravan parks and camping parks (708)

G 5.1   Introduction, purpose and principles

The 17th Edition has two separate sections concerning caravans, Section 708 covering caravan parks (and including camping parks) and a new Section 721 covering caravans and motor caravans. The latter, Section 721, is not included within this book as it is concerned with the internal wiring within caravans.

Section 708 does not apply to installations for mobile homes and similar and its scope is for 'caravan parks/camping parks' as defined in the Standard, as follows:

Caravan park/camping park: Area of land that contains two or more caravan pitches and/or tents.

Whereas a caravan is defined as:

Caravan: A trailer leisure accommodation vehicle, used for touring, designed to meet the requirements for the construction and use of road vehicles.

Caravan parks are laid out with 'caravan pitches' in the form of spaces to accommodate the caravans, and the layout will depend on the shape, topography and nature of the site. Most sites are laid out with the caravans in a regular arrangement to maximize the space, as parks are commercial enterprises. Others have trees, etc. and other features that make the park attractive and these need to be accommodated.

Caravans are for touring, and can travel throughout Europe, and farther, so their electrical installation may have to accommodate the different electrical practices of various countries, thus Section 721 requires the provision of a 30 mA RCD to the requirements of Regulation 415.1.1 to be provided within UK caravans.

G 5.2 Requirements and guidance

The requirements of BS 7671 are summarized in Table G 5.1.

The park provides an electricity supply at each pitch, and generally these supplies are provided by the use of a 16 A SP&N 'industrial' socket outlet to BS EN 60309–2. Some larger supplies may exist for very large caravans, but generally these would not be touring types.

As usual with special locations the protective measures of obstacles, placing out of reach, non-conducting location and earth-free equipotential bonding are not allowed as the parks and caravans are used by ordinary persons, and the installations cannot be under effective supervision at all times.

The electricity supply to a caravan park must be a TN-S supply to the caravan pitches. A TN-C-S supply is not acceptable as the common problems with a broken or lost neutral could make the metal frame and skin of a caravan live with respect to the general mass of earth, and this would be potentially dangerous for anyone

Table G 5.1 Caravan and camping park requirements.

Requirement	Regulation
TN-C-S arrangements are not permitted	708.411.4
Equipment to be minimum of IP 44	708.512.2
Overhead cables to be minimum of 6 m where vehicles are used and 3.5 m elsewhere	708.521.1.2
Supply point at pitch is to be a maximum of 20 m from caravan inlet	708.530.3
Pitch socket outlets to be provided with at least one outlet, minimum 16 A, IP44, BS EN 60529 type and located between 0.5 and 1.5 m high	708.553.1.8 708.553.1.9 708.553.1.10 708.553.1.11
Each pitch socket outlet to be provided with overcurrent protection and 30 mA RCD	708.553.1.12 708.553.1.13
Obstacles and Out Of Reach are not permitted	708.410.3.5
Non-conducting location and earth-free local equipotential bonding are not permitted	708.410.3.6

touching earth and the caravan at the same time. A TN-C-S system is allowable, however, in fixed buildings on the park, offices, shops, clubhouse, stores etc. and these can be treated as any other building.

Two common methods are used for installations with supplies with a TN-C-S PME incomer. The first is to break the PME at the distribution board that supplies the caravan pitches. The second, more popular, method is to break the PME supply at the pitch outlets. In both cases 'breaking' is normally carried out by terminating the earth cable sheath or similar on an insulated enclosure.

As parks are usually in exposed locations, particular care must be taken in selecting wiring systems and equipment, and further mechanical protection may be required for some applications. Generally pitch socket outlets should be IP44 if they are enclosed or protected from the elements, and higher IP ratings will be necessary if they are exposed. Impact protection must be considered, and it may be necessary to protect exposed cables with conduit or steel channel. The 17th Edition advises mechanical stress protection to AG3 (high severity) of Appendix 5 of BS 7671 but this does not give any practical advice as to what would be required.

Cables should be run round the site underground, but this may not be possible in all locations due to soil depth. The recommended minimum depth of cable burial is 600 mm, but where this cannot be achieved, mechanical protection must be provided if there is any likelihood of cable damage. Cables should be run round the edge of pitch locations or along road verges to reduce the possibility of damage caused by digging in pitch areas, or pegs or spikes used for awnings or tents.

Overhead conductors should only be considered where buried cables are not practical, and they should be high enough over roadways, or kept away from traffic routes to prevent damage from tall vehicles or caravans, as well as aerials, chimneys and other projections that may be on caravans. Overhead conductor supporting poles should also be located away from vehicle routes to avoid possible impact and damage.

Each caravan or tent pitch should be supplied with a conveniently located socket outlet to allow the caravan to connect easily and safely with its standard flexible connector.

In regularly laid-out parks, each group of perhaps four pitches would have a power supply location centrally located, but with irregularly laid-out parks this will not be so. Each socket outlet shall have individual overcurrent protection and a 30 mA RCD.

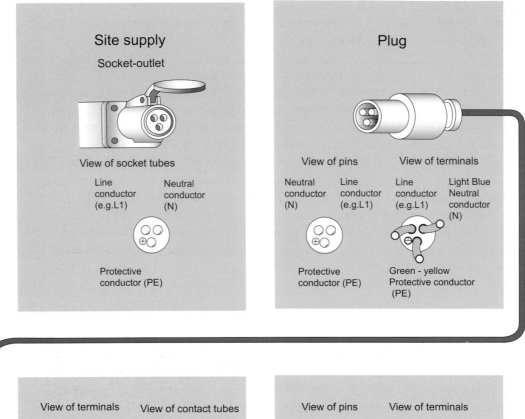

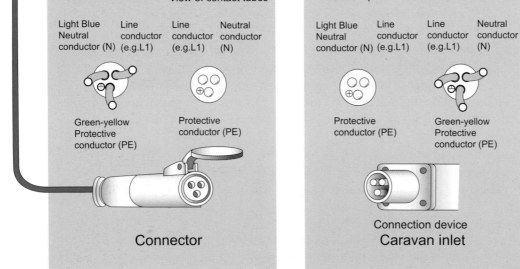

Figure G 5.1 Caravan pitch outlet, extension lead and inlet.

The 16th Edition of BS 7671 allowed up to three socket outlets to be supplied from one circuit and RCD, but the 17th Edition now requires each outlet to have its own RCD, to prevent interruption to other supplies by a faulty caravan. RCDs are to comply with the requirements of Regulation 415.1.1 and must be multi-pole.

Socket outlets should be fixed between 500 mm and 1500 mm from the ground to allow easy access and prevent damage from equipment or the elements. In cases where flooding may be expected, this height should be increased. The 16th Edition allowed socket outlets to be between 800 mm and 1500 mm from the ground, but this change in minimum dimension does not make installations carried out to the 16th Edition in any way unsafe, and the Regulations are not retrospective. This must be remembered during safety inspections. The further provision of RCDs in the 17th Edition has the same effect and does not affect safety.

Figure G 5.1 is adapted from a diagram in BS 7671: 2008 and shows a pitch outlet, extension lead and caravan inlet.

G 6 Exhibitions, shows and stands (711)

© Masterfile (Royalty-Free Division)

G 6.1 Introduction and risks

Section 711 of the special locations requirements applies to temporary exhibitions and shows, and the stands and exhibition equipment used in them. The section should not be applied to permanent exhibitions and shows, or to the permanent fixed installations of exhibition halls or show grounds that accommodate the exhibition. Vehicles or caravans that are used as prefabricated exhibition stands are included in this section.

The major risks are unsafe electrical work carried out by people who are not competent, giving rise to danger of fire and electric shock, and most exhibition and show organizers will require an electrical installation certificate for each stand's electrical installation before they will allow its connection to the local supply. Organizers also usually require the inspection and testing and certification of all electrical equipment (e.g. lighting, sound systems etc.) that is to be used. Major events are set up by professional engineering staff, and the use of pre-engineered stage components and systems is well established. Many smaller shows, however, may be provided for by local electrical contractors, and in very small shows the DIYer may be tempted to do his own work, with all the problems that may provide!

The environmental conditions must be properly assessed, and can range from agriculture shows in open fields, with attendant livestock, to exhibitions in regularly used purpose-built halls. The protective measures of obstacles or placing out of reach are not permitted, as they cannot be successfully managed, and non-conductive locations or earth-free local equipotential bonding are not permitted, as they are not under the continuous control of a competent person.

Generally, electrical systems should be as simple and robust as possible, and modular installation systems with plug and socket outlet connections have been developed to go with prefabricated modular stand constructions.

G 6.2 Requirements and guidance

The requirements of 711 are summarized in Table G 6.1.

Regulation 711.521 requires the selection of cables assuming the installation contains no fire alarm system in the building accommodating the exhibition or show. This requirement is not really practical as (apart from small private buildings such as school halls) all buildings open to the public in the UK will be subject to prior inspection by the local authority for licensing, and adequate fire alarm and emergency escape provisions are required for licensing. Even if this is not the case

Table G 6.1 Requirements for exhibitions, shows and stands.

Requirement	Regulation	Notes
A cable supplying a temporary exhibition, show or stand shall be additionally protected with a 300 mA RCD and shall discriminate with final circuits	711.410.3.4	For RCD discrimination see Section D 7.3
Obstacles and Out Of Reach not permitted Non-conducting location and earth-free local equipotential bonding not permitted	711.410.3.5 and 711.410.3.6	
Structural metal of stand, caravan, wagon or container to be main bonded	711.411.3.1.2	
Socket-outlet circuits up to 32 A and lighting circuits to have 30 mA RCD (excludes emergency lighting)	711.411.3.3	
Where SELV and PELV are used, basic protection shall be provided	711.414.4.5	
Cables shall be at least 1.5 mm²	711.52	
Butyl flex not to be used where no fire alarm is present	711.52	See text following table
Joints in cables shall only be made for connection into circuits	711.526.1	In-line joints should not be used
Joint system to be at least IPX4	711.526.1	
Separate units require isolators	711.537.2.3	
All motors shall have isolators located adjacent to them	711.55.4.1	
Adequate quantity of socket outlets shall be provided	711.55.7	
A visible switch shall be provided for signs, lamps or exhibit circuits	711.559.4.7	
Installation shall be inspected and tested after each assembly	711.6	

(as explained in the selection of cables section under Section D 5.4.1), the only BS EN cables that do not comply with BS EN 60332–1–2 are butyl flexes.

All supply cables are required to be protected against damage by an RCD with a rated residual operating current that does not exceed 300 mA, and all final circuits on the stand are to be protected by 30 mA RCD. This includes socket outlets rated up to and including 32 A, BS 7671, however, excludes emergency lighting circuits, but generally emergency lighting will be provided elsewhere in a building or show as part of the overall emergency escape policy and will not be a part of the stand (unless it is a very large stand with internal rooms or partitions), and emergency lighting will usually be of the self-contained type, so this exclusion is not relevant.

Installation methods must take into account the access of members of the public in relatively large numbers, and the fact that these persons generally will be more interested in the exhibition or show than anything else, and so cables must be run in safe locations away from damage, trip hazards and wear from users (e.g. foot or vehicle traffic), and armoured cables or other protection should be used where there is expected to be any risk of damage.

It is desirable that switchgear and control gear is not accessible to the general public and should be enclosed in lockable cabinets or similar.

Most exhibition halls and shows have specific licensing and inspection requirements or rules for exhibitors, e.g. the Amusement Devices Inspection Procedures Scheme (ADIPS). This was introduced by the industry with HSE support, and provides for third party safety inspections for entertainment and amusement devices and rides. Current approval documentation from such an inspection may need to be shown to the organizers, along with electrical installation and electrical equipment inspection and testing certification.

G 7   Solar photovoltaic (PV) power supply systems (712)

G 7.1   Introduction, principles and terminology

BS 7671 includes Section 712, which, to give it its full title, is 'Solar Photovoltaic (PV) Power Supply Systems'. The purpose for inclusion of this new section in BS 7671 is only concerned with safety.

In this section of the book, terms used to describe the subject are PV cells (individual cells), PV strings (a circuit arrangement of PV cells), PV arrays (a general term for a collection of cells, possibly comprising a number of PV strings) and PV system (the cells or arrays, their control and connection); the terms are generally expanded upon and illustrated in this section.

A PV system is a collection of interconnected PV cells that turn sunlight directly into electrical energy, and consequently needs to be installed outside (usually at roof level); so any external electrical work has to be suitable for the environment and correctly IP rated. PV arrays and equipment must conform with the relevant equipment Standards, i.e. BS EN 61215.

PV arrays are just one of several sources of sustainable energy (e.g. wind turbines), and all have similar connection requirements to run in parallel with the Regional Electricity Companies' mains supply. The Department of Trade and Industry (DTI) has published a guidance document on the installation and connection of such sources (Photovoltaics in Buildings, 2nd Edition printed 2006 {URN 06/1972}), and this provides detailed information on installation requirements.

PV arrays produce electricity at a voltage dependent upon the physical characteristics of the array and its construction, and the series and parallel interconnection of PV cells within an array. These are all set during the manufacturing process and the electrical contractor only has to install the system to the manufacturer's details.

As the PV array is a source of energy, it is effectively live at all times and cannot be isolated (like a battery). The array output voltage is dependent on the array construction and the load current, and needs to be provided by the manufacturer's designer. When on load, like any source, the terminal voltage can fall to zero at high load, but will recover when the load is removed.

Figure G 7.1 shows the no-load current-voltage characteristic of a typical PV cell, and Figure G 7.2 shows a typical loaded power–voltage characteristic.

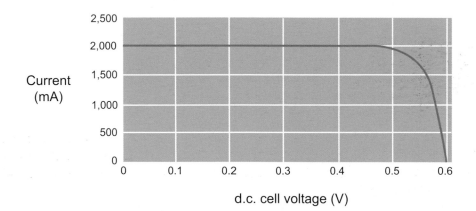

Figure G 7.1 Typical PV cell no-load current-voltage characteristic.

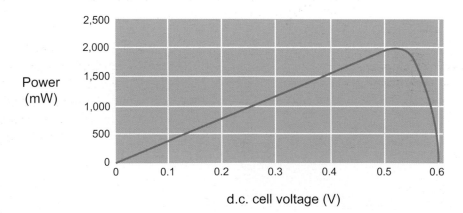

Figure G 7.2 Typical PV cell on-load power–voltage characteristic.

Table G 7.1 PV system requirements.

Requirement	Regulation
PV a.c. supply to be connected to supply side of protective device	712.411.3.2.1.1
RCD required and shall be type B to IEC 60775[1]	712.411.3.2.1.2
Class II is preferred on d.c. side of cell	712.412
Overload protection may be omitted from the PV array string if the cable current-carrying capacity is rated to at least 1.25 of the short circuit current. Short circuit current protection must be provided at connection to the mains	712.433.1 712.434.1
Isolation for maintenance on d.c. and a.c. sides to be provided	712.537.2.1.1
All junction boxes to carry label warning about energization after loss of mains	712.537.2.2.5.1
Protective bonding conductors to be run in close contact with d.c. and a.c. PV system cables	712.54.

[1] RCD not required where the PV construction is not able to feed the d.c. fault.

G 7.2   Requirements

The requirements of BS 7671 are summarized in Table G 7.1.

G 7.3   Notes and guidance

It is required to provide an isolator in a suitable location on the d.c. output from the array, and it must be noted that the cables from the array to this isolator will always be live, and must be of a suitable construction to withstand thermal and mechanical damage.

The general method of installation is outlined in Figure G 7.3, but PV arrays are at present specially manufactured items, usually supplied as a complete system with the converter and specialist installation in a 'package'.

There are mandatory requirements concerning parallel connection of 'generators' before installations can be interconnected with the supply network. Section 551 of BS 7671 has general regulations on these connections (see 551 and Section D 9 of this book). The permission of the local distribution network operator (DNO, formerly known as REC) must be obtained, and this is a relatively simple, formal process for small domestic systems up to 5 kW, but the process becomes considerably more complex for larger commercial projects.

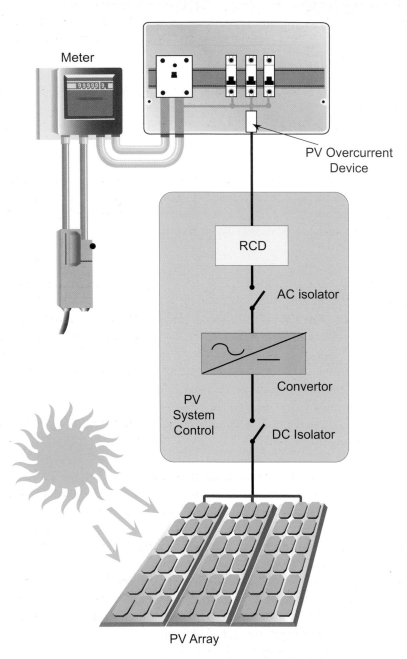

Meter

PV Overcurrent Device

RCD

AC isolator

Convertor

PV System Control

DC Isolator

PV Array

Figure G 7.3 Typical PV system, showing array and control and isolation.

To aid designers and installers the DNOs have issued *Engineering Recommendations* to assist in designing and specifying systems correctly for their approval. Recommendation G83/1 is for PV systems up to 5 kW, and G59/1 covers PV and other systems above 5 kW. These documents can be purchased from the Energy Networks Association.

Protection by the use of Class II (double insulated) or equivalent insulation is advised for the d.c. system from the array, but it is permitted to earth the d.c. side at one point if there is at least separation with the use of basic insulation between the d.c. and a.c. sides.

All equipment used on the d.c. side must be suitable for d.c. voltages and currents, equipment approved to normal a.c. standards will not be suitable (especially switchgear), and the designer should clarify standards and performance requirements with equipment manufacturers.

PV arrays must be installed by competent persons to an approved design, and planning and building regulations approvals may be required. As they are fixed to the outside of a building they will be subject to all environmental conditions including storms and high winds, so the construction and weather sealing must be sound. The installation must also be accessible for any repair or maintenance, although with no moving parts this should be minimal. The arrays are to be installed to allow adequate ventilation to prevent any heat build-up, especially to electrical equipment and components.

Depending on the design of the arrays, overcurrent protection may not be necessary on the d.c. side cables as there is a limit to the current output of the array.

$I_{SC\ STC}$ is the short circuit current of a PV module or array under standard test conditions, and $V_{OC\ STC}$ is the open circuit voltage under standard test conditions of an unloaded PV module or array on the d.c. side of the system. If the continuous current-carrying capacity of any d.c. cable at any point is at least 1.25 $I_{SC\ STC}$ with any de-rating factors taken into account, overcurrent protection is not required.

Short circuit protection, however, must be provided at the connection to the a.c. mains for the a.c. side supply cable, in accordance with the normal requirements of Part 4 of BS 7671, to prevent damage from an a.c. side fault (see Figure G 7.3).

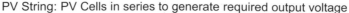

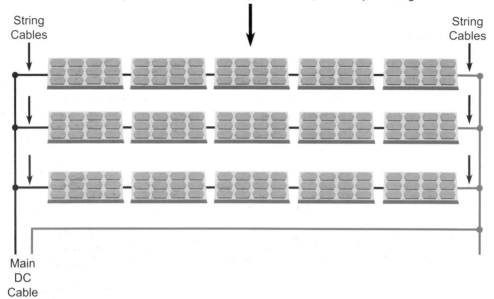

Figure G 7.4 PV string arrangement.

Cables shall be of a type selected to be suitable for their environment and conditions of use, and suitable for the expected equipment temperatures (DTI guidance suggests 80°C rated cables should be considered). Cables must also be installed to minimize the possible risks of damage and faults, and further protection or armoured cabling may be necessary.

Plug and socket connectors, specific to PV systems, are commonly fitted to module cables by the manufacturer. Such connectors provide a secure, durable and effective electrical contact. They also simplify and increase the safety of installation works. They are recommended in particular for any installation being performed by a non-PV specialist, for example, a PV array being installed by a roofer.

G 8 Mobile or transportable units (717)

G 8.1 Scope and application

New for the 17th Edition, this is perhaps the most unusual section of the special locations, as it covers a wide range of units from specialist broadcast vehicles to simple skid-mounted units. Generally all such units are prefabricated to specific user specifications and technical standards, and the general electrical consulting engineer or electrical contractor will have little to do with their construction.

You may now ask why this section has been included within the book. Although the design of the units is carried out by manufacturers, questions concerning connection to the units, or their inspection, are ones that are frequently put to electrical contractors.

The scope of the section states that the 'units' are intended to mean either a mobile vehicle unit or a transportable unit; for example, a container.

It cites examples as: technical and facilities vehicles for the entertainment industry, medical services, advertising, fire fighting, workshops/offices and transportable catering units. It then lists exclusions such as generating sets, marinas and pleasure craft, mobile machinery to BS EN 60204–1 (Safety of Machinery).

Thus the whole application of Section 717 is confused, as it could be applied to a collection of temporary site buildings, or Portakabins® used as wards for the local NHS hospital.

Whilst some of the measures in this section would be suitable, such constructions would not normally be expected to be 'special locations' and the normal installation requirements of Parts 1 to 5 of BS 7671 would be adequate. This book, therefore, limits itself to specialist mobile units; even this is not easy to define but would include, for example, outside broadcast mobile vehicles.

To throw some advice at those of you who are a little lost, perhaps a solution would be to consider the requirements given in Section G 8.2; there is not a vast quantity and they are not particularly difficult to apply. One option open therefore would be, if in doubt, to apply the requirements anyway.

Table G 8.1 Mobile or transportable units system requirements

Requirement	Regulation
Automatic disconnection shall be by an RCD	717.411.1
Socket outlets to be 30 mA protected (unless they are SELV, PELV or by electrical separation)	717.415
Accessible metal parts like the chassis shall be main bonded, using finely stranded conductors	717.411.3.1.2
TN-C-S (PME) shall not be used	717.411.4
Unit supply cables to be flexible to BS 7919 or equivalent, minimum 2.5 mm^2 and shall enter the unit via an insulating sleeve or enclosure	717.52.1
Unit supply connectors to be BS EN 60309–2, insulated to IP44	717.55.1
Electrical equipment not to be run in gas storage area except gas-supply control	717.528.3.5
Unit to have an electrical rating plate, indicating earthing arrangement, voltage, phases and maximum power	717.514
Obstacles and Out Of Reach are not permitted	717.417
Non-conducting location or earth free local equipotential bonding are not permitted	717.418
Alternatively, an IT supply can be used, with insulation monitoring, and RCD to disconnect if the isolating transformer fails	717.411.6.2

G 8.2   Requirements

The requirements of Section 717 are summarized in Table G 8.1; some of these requirements are items that should or can only be undertaken at the manufacturing stage.

G 8.3   Notes and guidance

Generally for such units a TN-C-S supply system is not allowable, due to the possible dangers associated with a broken or disconnected neutral making the earth a part of the circuit, and the need to properly control and maintain such a supply system; not easily practical in most 'temporary' locations.

The integrity of earthing and bonding connections is obviously of considerable concern in such mobile units, where vibration can be a considerable problem, and the Regulations require that all accessible metalwork parts of the unit are to be connected through a finely stranded protective bonding conductor to the main earthing terminal of the unit. A note to Regulation 717.411.3.1.2 suggests that cable types H05V-K and H07V-K to BS 6004 are appropriate stranded cable types. Generally, however, as discussed previously, such mobile or transportable units are designed by specialists for special applications and any earthing and bonding connections should be carried out by the contractor to the instructions of the equipment designer.

Generally, the protective measure of automatic disconnection of the supply is achieved by the use of an RCD.

Some specialist units such as broadcast vehicles do not use RCDs on their earthing system, as the sudden interruption of supply could not be tolerated. These installations are always under the control of competent persons during operation, and the earth current is closely monitored.

Small units may be operated more practically as an IT system (isolated from earth). The 'first fault' will not cause operation of any protective device, but the unit should be constructed and operated to minimize the likelihood of any such fault.

BS 7671 however, allows IT systems to be provided only by either the use of an insulation monitoring device or by the use of simple electrical separation. Simple separation must itself provide for the use of an insulation monitoring device providing automatic disconnection at first fault, or the use of a 30 mA RCD complying with Regulation 415.1.1 of BS 7671 (30 mA).

As with all special locations there is increased risk of danger if the installation is not properly designed, constructed and commissioned, and then properly operated and maintained by competent persons. As such units will usually be used outside, and be subject to continual movement and handling, the electrical equipment must be suitably rated, environmentally protected, of good quality and be to relevant British or equivalent Standards.

Cables for internal wiring of units may be PVC insulated single-core to BS 6004 installed in conduit; but if there is any possibility of vibration, all-insulated conductors, or even flexible stranded conductors, should be used.

Cables for connecting the unit to the supply must be of flexible type, and suitable for their environment and method of use. Such cables should be protected if there is any likelihood of damage (e.g. foot or vehicle traffic) and protected from any abrasion. BS 7671 recommends that flexible H05RN-F or H07RN-F cable of a minimum 2.5 mm^2 cross-sectional area be used, but this may need to be increased for reasons of voltage drop.

Plugs and socket outlets are to be industrial type to BS EN 60309-2, insulated construction only, and be suitably IP rated to a minimum of IP44 outside. The plug top shall be on the unit.

Finally, Section 717 includes two drawings of internal wiring arrangements for mobile units; these diagrams are most confusing and do not greatly extend the reader's knowledge.

G 9 Floor and ceiling heating systems (753)

G 9.1 Introduction

Section 753 'Floor and Ceiling Heating Systems' is new to BS 7671: 2008. Readers may wonder whether this is Section 753 or 53rd in the series of Part 7s. The answer is that within IEC, there are many Part 7s either drafted or planned.

Electric floor and ceiling heating systems are mainly used in domestic and similar installations and are usually buried in the building fabric.

The main risks of such systems are overheating and physical damage after installation, as the systems are not visible but buried at a shallow depth.

G 9.2 Requirements

Again the requirements have been summarized in a Table, G 9.1.

Table G 9.1 Floor and ceiling heating requirements.

Requirement	Regulation
Automatic disconnection shall be by a 30 mA RCD	753.411.3.2
Heating units (cables, tape or panels) that are not manufactured with a conductive covering or mesh shall be protected with a metal mesh or grid with spacing of not more than 30 mm	753.411.3.2
A maximum temperature of 80°C operating temperature shall be achieved by design, installation or by sensing. Floor areas shall have a lower temperature for skin comfort (say 35°C)	753.424.1.1 753.423
Heating cables shall be to BS 6351 and flexible sheet panels to BS EN 60335–2-96	753.511
Heating units (cables, type of panels) for floor installation shall be to IPX7	753.512.2.5
Information shall be given to the user of the installation (see Section G 9.3)	753.514

G 9.3   Notes and guidance

Systems are to be protected by a 30 mA RCD. If heaters do not have an integral conductive covering or sheath (an exposed-conductive-part), a conductive covering with a spacing of not more than 30 mm (e.g. a grid or mesh) is to be provided at installation as an exposed-conductive-part over the heating system. See also Regulation 701.753 for bathrooms.

A grid or mesh will not provide full mechanical protection against physical damage from nails, screws and the like, but is intended to provide a conductive path to earth, and not physical protection in its own right.

A Class II heating system (or of equivalent construction) is also to be provided with additional protection by using a 30 mA RCD.

Unfortunately, RCDs can be subject to unwanted tripping, and so the load on each RCD should be limited to ensure that any leakage current is limited to levels that will not cause unwanted operation of the RCD. A maximum circuit load of 7.5 kW for single-phase loads or 13 kW for three-phase loads are suggested in Section 753.

All such heating systems must be installed in accordance with manufacturers' instructions, and the possible temperature of connections and the local ambient temperatures must be considered in any installation.

It is required that the installer (or designer) provides a record drawing for each heating system installed showing its location, area, rating details and, indeed, far more information than will ever actually be provided, in most cases.

The complete list of information to be provided is given below.

A description of the heating system shall be provided by the installer of the heating system to the owner of the building or his/her agent upon completion of the installation.

The description shall contain at least the following information:

a) Description of the construction of the heating system, which must include the installation depth of the heating units.
b) Location diagram with information concerning:
    ● the distribution of the heating circuits and their rated power;

- the position of the heating units in each room;
- conditions which have been taken into account when installing the heating units, for example, heating-free areas, complementary heating zones, unheated areas for fixing means penetrating into the floor covering.

c) Data on the control equipment used, with relevant circuit diagrams and the dimensioned position of floor temperature and weather conditions sensors, if any.

d) Data on the type of heating units and their maximum operating temperature.

The installer shall inform the owner that the description of the heating system includes all necessary information, for example, for repair work.

The installer shall provide the owner with a description of the heating system including all necessary information, for example, to permit repair work. In addition, the installer shall provide instructions for use of the heating installation.

The designer/installer of the heating system shall hand over an appropriate number of instructions for use to the owner or his/her agent upon completion. One copy of the instructions for use shall be permanently fixed in or near each relevant distribution board.

The instructions for use shall include at least the following information:

a) Description of the heating system and its function.

b) Operation of the heating installation in the first heating period in the case of a new building, for example, regarding drying out.

c) Operation of the control equipment for the heating system in the dwelling area and the complementary heating zones as well, if any.

d) Information on restrictions on placing of furniture or similar. Information provided to the owner shall cover the restrictions, if any, including:
   - whether additional floor coverings are permitted, for example, carpets with a thickness of >10 mm may lead to higher floor temperatures which can adversely affect the performance of the heating system;
   - where pieces of furniture solidly covering the floor and/or built-in cupboards may be placed on heating-free areas;
   - where furniture, such as carpets, seating and rest furniture with pelmets, which in part do not solidly cover the floor, may not be placed in complementary heating zones, if any.

e) Information on restrictions on placing of furniture or similar.

f) In the case of ceiling heating systems, restrictions regarding the height of furniture. Cupboards of room height may be placed only below the area of ceiling where no heating elements are installed.

g) Dimensioned position of complementary heating zones and placing areas.

h) Statement that, in the case of thermal floor and ceiling heating systems, no fixing shall be made into the floor and ceiling respectively. Excluded from this requirement are unheated areas. Alternatives shall be given, where applicable.

References

British Standards Institution (1958)
BS 3036: 1958 *Specification. Semi-enclosed electric fuses (rating up to 100 amperes and 240 volts to earth)*
London: BSI

British Standards Institution (1971)
BS 1361: 1971 *Specification for cartridge fuses for a.c. circuits in domestic and similar premises*
London: BSI

British Standards Institution (1979)
PD 6484: 1979 *Commentary on corrosion at bimetallic contacts and its alleviation*
London: BSI

British Standards Institution (1988)
BS 88 part 1: 1988 *Cartridge fuses for voltages up to and including 1000 V a.c. and 1500 V d.c.*
London: BSI

British Standards Institution (1998)
BS 7430: 1998 *Code of Practice for Earthing*
London: BSI

British Standards Institution (2003)
BS EN 60898 part 1: 2003 *Specification for circuit-breakers for overcurrent protection for household and similar installations*
London: BSI

British Standards Institution (2008)
BS 7671: 2008 *Requirements for Electrical Installations – IEE Wiring Regulations Seventeenth Edition*
London: BSI

Coates, M. and Jenkins, B. D. (2003)
Electrical Installation Calculations 3rd ed.
Oxford: Blackwell Science Ltd

Kaplan, S. M. (2004)
Wiley Electrical And Electronics Engineering Dictionary
Hoboken: John Wiley & Sons

Appendices

Appendices 1–5 are contained within this chapter; Appendices 10–17 are available via the Companion Website (see below for details).

1 Standards and bibliography
2 Popular cables current rating tables from BS 7671: 2008 Appendix 4 4E1A 4E4A and 4D5A (samples only)
3 Limiting earth fault loop impedance tables from BS 7671: 2008
4 Cable data-resistance, impedance and $R_1 + R_2$ values
5 Fuse I^2t characteristics

The following appendices are included on the Companion Website available at http://www.wiley.com/go/eca_wiringregulations

10 Example cable sizing calculations
11 Impedances of conduits and trunking
12 Earth electrodes and earth electrode testing
13 Notes on Out of Reach/Obstacles/Non-conducting Location/Earth-free Local Equipotential Bonding (see 4.6.1)
14 Additional 'occasional' tests that may be required
15 Notes on periodic inspection and testing
16 Electrical Research Association (ERA) report on armoured cables with external cpcs
17 A4 sample versions of ECA BS 7671: 2008 certificates and forms
18 Building Standards in Scotland

Appendix 1 – Standards and bibliography

Standards

The following list of standards may be used or encountered when undertaking electrical installation work. This list is not expected to be used other than as a look-up table when a standard number is quoted and only a fraction of these standards were used when writing this book. The following table does not include the date of the standard, merely the title.

The BSI's website for 'standards on line' is an excellent database and tool for searching standards. It is available to search free of charge for viewing the title of standards and is accessed via

http://www.bsonline.bsi-global.com

Standard no.	Title
BS 31	Specification. Steel conduit and fittings for electrical wiring
BS 67	Specification for ceiling roses
BS 88[1]	Cartridge fuses for voltages up to and including 1000 V a.c. and 1500 V d.c.
BS 196	Specification for protected-type non-reversible plugs, socket-outlets, cable couplers and appliance couplers with earthing contacts for single-phase a.c. circuits up to 250 volts
BS 476[1]	Fire tests on building materials and structure
BS 546	Specification. Two-pole and earthing pin plugs, socket outlets and socket outlet adaptors
BS 559	Specification for electric signs and high voltage luminous discharge tube installations
BS 646	Specification. Cartridge fuse links (rated up to 5 amperes) for a.c. and d.c. service
BS 731	Flexible steel conduit for cable protection and flexible steel tubing to enclose flexible drives
BS 731–1	Flexible steel conduit and adaptors for the protection of electric cables
BS 951	Specification for clamps for earthing and bonding purposes
BS 1361	Specification for cartridge fuses for a.c. circuits in domestic and similar premises
BS 1362	Specification for general purpose fuse links for domestic and similar purposes (primarily for use in plugs)
BS 1363[1]	13 A plugs, socket outlets, connection units and adaptors
BS 3036	Specification. Semi-enclosed electric fuses (rating up to 100 amperes and 240 volts to earth)
BS 3535[1]	Isolating transformers and safety isolating transformers; to be jointly read with the BS EN 61558 series of standards
BS 3676	Switches for household and similar fixed electrical installations
BS 4066[1]	Tests on electric cables under fire conditions

Standard no.	Title
BS 4363	Specification for distribution assemblies for electricity supplies for construction and building sites
BS 4444	Guide to electrical earth monitoring and protective conductor proving
BS 4553[1]	Specification for 600/1000 V single-phase split concentric electric cables
BS 4568	Specification for steel conduit and fittings with metric threads of ISO form for electrical installations
BS 4568	Steel conduit, bends and couplers
BS 4573	Specification for 2-pin reversible plugs and shaver socket outlets
BS 4607	Non-metallic conduits and fittings for electrical installations
BS 4678[1]	Cable trunking
BS 4727	Glossary of electrotechnical, power, telecommunications, electronics, lighting and colour terms
BS 5042	Specification for bayonet lampholders (replaced by BS EN 61184)
BS 5266	Emergency lighting
BS 5345	Electrical apparatus for explosive gas atmospheres (replaced in part by BS EN 60079) and BS EN 50014: 1998 Electrical apparatus for potentially explosive atmospheres
BS 5467	Specification for 600/1000 V and 1900/3300 V armoured electric cables having thermosetting insulation
BS 5518	Specification for electronic variable control switches (dimmer switches) for tungsten filament lighting
BS 5655[1]	Lifts and service lifts
BS 5733	Specification for general requirements for electrical accessories
BS 5839[1]	Fire detection and alarm systems for buildings
BS 6004	2000 Electric cables. PVC insulated, non-armoured cables for voltages up to and including 450/750 V, for electric power, lighting and internal wiring
BS 6141	Specification for insulated cables and flexible cords for use in high temperature zones
BS 6207[1]	Mineral insulated cables with a rated voltage not exceeding 750 V
BS 6346	Specification for 600/1000 V and 1900/3000 V armoured cables having PVC insulation
BS 6351	Electric surface heating
BS 6351[1]	Specification for electric surface heating devices etc.
BS 6423	Code of practice for maintenance of electrical switchgear and control gear for voltages up to and including 1 kV
BS 6458[1]	Fire hazard testing for electrotechnical products
BS 6500	Electric cables. Flexible cords rated up to 300/500 V, for use with appliances and equipment intended for domestic, office and similar environments
BS 6651	Code of practice for protection of structures against lightning
BS 6701	Code of practice for installation of apparatus intended for connection to certain telecommunications systems

Standard no.	Title
BS 6724	Specification for 600/1000 V and 1900/3300 V armoured cables having thermosetting insulation and low emission of smoke and corrosive gases when affected by fire
BS 6883	Elastomer insulated cables for fixed wiring in ships and on mobile and fixed offshore units
BS 6907	Electrical installations for open cast mines and quarries
BS 6972	Specification for general requirements for luminaire supporting couplers for domestic, light industrial and commercial use
BS 6991	Specification for 6/10 amp two pole weather-resistant couplers for household, commercial and light industrial equipment
BS 7001	Specification for interchangeability and safety of a standardized luminaire supporting coupler (to be read in conjunction with BS 6972)
BS 7071	Specification for portable residual current devices
BS 7211	Specification for thermosetting insulated cables with low emission of smoke and corrosive gases when affected by fire
BS 7288	Specification for socket outlets incorporating residual current devices (SRCDs)
BS 7375	Code of practice for distribution of electricity on construction and building sites
BS 7430	Code of practice for earthing
BS 7454	Method for calculation of thermally permissible short-circuit currents taking into account non-adiabatic heating effects
BS 7629[1]	Specification for 300/500 V fire-resistant electric cables having low emission of smoke and corrosive gases when affected by fire
BS 7697	Nominal voltages for low voltage public electricity supply systems
BS 7769[1]	Electric cables, Calculation of current rating
BS 7822[1]	Insulation coordination for equipment within low voltage systems
BS 7846	Electric cables. 600/1000 V armoured fire resistant electric cables having thermosetting insulation and low emission of smoke and corrosive gases when affected by fire
BS 7889	Specification for 600/1000 V single-core unarmoured electric cables having thermosetting insulation
BS 7895	Specification for bayonet lampholders with enhanced safety
BS 7919	Electric cables. Flexible cables rated up to 450/750 V, for use with appliances and equipment intended for industrial and similar environments
BS 8436	300/500 V screened electric cables having low emission of smoke and corrosive gases when affected by fire, for use in thin partitions and building voids
BS EN 1648[1]	Leisure accommodation vehicles. 12 V direct current extra low voltage electrical installations
BS EN 50110[1]	Operation of electrical installations
BS EN 50014	Electrical apparatus for potentially explosive atmospheres
BS EN 50081	Electromagnetic compatibility. Generic emission standard

Standard no.	Title
BS EN 50082	Electromagnetic compatibility. Generic immunity standard. (Partially replaced by BS EN 61000–6-2: 1999)
BS EN 50085[1]	Specification for cable trunking and ducting systems for electrical installations
BS EN 50086[1]	Specification for conduit systems for electrical installations
BS EN 50265	Common test methods for cables under fire conditions. Test for resistance to vertical flame propagation for a single insulated conductor or cable
BS EN 50265[1]	Tests for resistance to vertical flame propagation for a single insulated conductor or cable
BS EN 50281	Electrical apparatus for use in the presence of combustible dust
BS EN 50310	Application of equipotential bonding and earthing in buildings with information technology equipment
BS EN 60079[1]	Electrical apparatus for potentially explosive gas atmospheres
BS EN 60238	Specification for Edison screw lampholders
BS EN 60269[1]	Low voltage fuses
BS EN 60309[1]	Plugs, socket outlets and couplers for industrial purposes.
BS EN 60335[1]	Household appliances (many parts, about a hundred)
BS EN 60423	Conduits for electrical purposes. Outside diameters of conduits for electrical installations and threads for conduits and fittings (replaces BS 6053)
BS EN 60439[1]	Specification for low voltage switchgear and control gear assemblies
BS EN 60445	Basic and safety principles for man-machine interface, marking and identification. Identification of equipment terminals and of terminations of certain designated conductors, including general rules for an alphanumeric system
BS EN 60446	Basic and safety principles for the man-machine interface, marking and identification. Identification of conductors by colours or numerals
BS EN 60529	Specification for degrees of protection provided by enclosures (IP code)
BS EN 60570	Electrical supply track systems for luminaires
BS EN 60598	Luminaires
BS EN 60598–2-24	Luminaires with limited surface temperature
BS EN 60617	Graphical symbols for diagrams
BS EN 60669[1]	Switches for household and similar fixed electrical equipment
BS EN 60702[1]	Mineral insulated cables and their terminations with a rated voltage not exceeding 750 V
BS EN 60742	Isolating transformers and safety isolating transformers
BS EN 60898	Specification for circuit-breakers for overcurrent protection for household and similar installations
BS EN 60947[1]	Specification for low voltage switchgear and control gear
BS EN 60947[1]	Switches, disconnectors, switch-disconnectors and fuse-combination units

Standard no.	Title
BS EN 60950	Specification for safety of information technology equipment including electrical business equipment
BS EN 61008[1]	Residual current operated circuit-breakers without integral overcurrent protection for household and similar uses (RCCBs)
BS EN 61009	Residual current operated circuit-breakers with integral overcurrent protection for household and similar uses (RCBOs)
BS EN 61011[1]	Electric fence energizers. Safety requirements for mains operated electric fence energizers
BS EN 61184	Bayonet lampholders
BS EN 61386	Conduit system for cable management. General requirements
Note 1: Contains various parts	

Bibliography and further reading
Readers may find the following publications useful as additional reading.

- IEE Guidance Notes 1 to 8 inclusive
- IEE Commentary on BS 7671: 2001
- Cooper Development Agency (CDA), various publications on harmonics, also mirrored to some extent by BSRIA. See websites for details:
  - www.cda.org.uk
  - www.bsria.co.uk

2 Appendix 2 – Popular cables: current rating tables from BS 7671: 2008 Appendix 4

Three tables are included here, BS 6004 PVC thermosetting flat twin and earth, XLPE thermosetting single-core and XLPE thermosetting armoured. These tables are not complete and have been added to make this book, particularly Chapter C, readable. For complete tables, please refer to BS 7671: 2008.

Table 4C5 Rating factors for groups of one or more circuits of single-core cables to be applied to reference current-carrying capacity for one circuit of single-core cables in free air – Reference Method F in Tables 4D1A to 4J4A.

Samples from Table 4 D5A 70°C thermoplastic insulated and sheathed flat cable with protective conductor (copper conductors).

Current-carrying capacity (A)

Ambient temperature: 30°C
Conductor operating temperature: 70°C

Conductor cross-sectional area	Reference Method 100 (above a plasterboard ceiling covered by thermal insulation not exceeding 100 mm in thickness)	Reference Method 101 (above a plasterboard ceiling covered by thermal insulation exceeding 100 mm in thickness)	Reference Method 102# (in a stud wall with thermal insulation with cable touching the inner wall surface)	Reference Method 103# (in a stud wall with thermal insulation with cable not touching the inner wall surface)	Reference Method C* (clipped direct)	Reference Method A* (enclosed in conduit in an insulated wall)	Voltage drop (per ampere per metre)
1	2	3	4	5	6	7	8
(mm²)	(A)	(A)	(A)	(A)	(A)	(A)	(mV/A/m)
1	13	10.5	13	8	16*	11.5	44
4	27	22	27	17.5	37	26	11
16	57	46	63	42.5	85	57	2.8

A*   For full installation method refer to Table 4A2 Installation Method 2 but for Twin flat and earth cable
B*   For full installation method refer to Table 4A2 Installation Method 20 but for Twin flat and earth cable
100# For full installation method refer to Table 4A2 Installation Method 100
101# For full installation method refer to Table 4A2 Installation Method 101
102# For full installation method refer to Table 4A2 Installation Method 102
103# For full installation method refer to Table 4A2 Installation Method 103
Wherever practicable, a cable is to be fixed in a position such that it will not be covered with thermal insulation. Regulation 523.7, BS 5803-5: Appendix C: Avoidance of overheating of electric cables, Building Regulations Approved Document B and Thermal insulation: avoiding risks, BR 262, BRE, 2001 refer.

Samples from Table 4 E1A Single-core 90°C thermosetting insulated cables, unarmoured, with or without sheath (copper conductors).

Current-carrying capacity (A)

Ambient temperature: 30°C
Conductor operating temperature: 90°C

Conductor cross-sectional area	(enclosed in conduit in thermally insulating wall, etc.)		(enclosed in conduit on a wall or in trunking, etc.)		(clipped direct)		(in free air or on a perforated cable tray etc horizontal or vertical, etc.) Touching			(in free air) Spaced by one cable diameter — 2 cables, single-phase a.c. or d.c. or 3 cables three-phase a.c. flat	
	2 cables, single-phase a.c. or d.c.	3 or 4 cables, three-phase a.c.	2 cables, single-phase a.c. or d.c.	3 or 4 cables, three-phase a.c.	2 cables, single-phase a.c. or d.c. flat and touching	3 or 4 cables, three-phase a.c. flat and touching or trefoil	2 cables, single-phase a.c. or d.c. flat	3 cables, three-phase a.c. flat	3 cables, three-phase a.c trefoil	Horizontal	Vertical
1	2	3	4	5	6	7	8	9	10	11	12
(mm²)	(A)	(A)	(A)	(A)	(A)	(A)	(A)	(A)	(A)	(A)	(A)
1.5	9	17	23	20	25	23	—	—	—	—	—
50	158	141	198	175	228	209	242	216	207	275	246
120	278	249	354	312	413	379	437	400	383	500	454
300	486	435	603	514	743	681	783	736	703	902	833

Samples from Table 4 E4A Multicore 90°C armoured thermosetting insulated cables (copper conductors).

Current-carrying capacity (A)

Air ambient temperature: 30°C
Ground ambient temperature: 20°C
Conductor operating temperature: 90°C

Conductor cross-sectional area	(clipped direct)		(in free air or on a perforated cable tray etc, horizontal or vertical)		(direct in ground or in ducting in ground, in or around buildings)	
1	2	3	4	5	6	7
	1 two-core cable, single-phase a.c. or d.c.	1 three- or 1 four-core cable, three-phase a.c.	1 two-core cable, single-phase a.c. or d.c.	1 three- or 1 four-core cable, three-phase a.c.	1 two-core cable, single-phase a.c. or d.c.	1 three- or 1 four-core cable, three-phase a.c.
(mm²)	(A)	(A)	(A)	(A)	(A)	(A)
6	62	53	66	56	53	44
25	146	124	152	131	116	96
50	219	187	228	197	164	135
185	515	441	539	463	343	281

3 Appendix 3 – Limiting earth fault loop impedance tables from BS 7671: 2008

Table 41.2 Maximum earth fault loop impedance (Z_s) for fuses, for 0.4 s disconnection time with U_0 of 230 V (see Regulation 411.4.6).

(a) General purpose (gG) fuses to BS 88–2.2 and BS 88–6					
Rating (A)	6	10	16	20	25	32
Z_s (Ω)	8.52	5.11	2.70	1.77	1.44	1.04

(b) Fuses to BS 1361			
Rating (A)	5	15	20	30
Z_s (Ω)	10.45	3.28	1.70	1.15

(c) Fuses to BS 3036				(d) Fuses to BS 1362		
Rating (A)	5	15	20	30	Rating (A)	3	13
Z_s (Ω)	9.58	2.55	1.77	1.09	Z_s (Ω)	16.4	2.42

NOTE: The circuit loop impedances given in the table should not be exceeded when the conductors are at their normal operating temperature. If the conductors are at a different temperature when tested, the reading should be adjusted accordingly.

Table 41.3 Maximum earth fault loop impedance (Z_s) for circuit breakers with U_0 of 230 V, for instantaneous operation giving compliance with the 0.4 s disconnection time of Regulation 411.3.2.2 and 5 s disconnection time of Regulation 411.3.2.3.

(a) Type B circuit breakers to BS EN 60898 and the overcurrent characteristics of RCBOs to BS EN 61009													
Rating (A)	3	6	10	16	20	25	32	40	50	63	80	100	125	In
Z_s (Ω)	15.33	7.67	4.60	2.87	2.30	1.84	1.44	1.15	0.92	0.73	0.57	0.46	0.37	46/In

(b) Type C circuit breakers to BS EN 60898 and the overcurrent characteristics of RCBOs to BS EN 61009													
Rating (A)		6	10	16	20	25	32	40	50	63	80	100	125	In
Z_s (Ω)		3.83	2.30	1.44	1.15	0.92	0.72	0.57	0.46	0.36	0.29	0.23	0.18	23/In

(c) Type D circuit breakers to BS EN 60898 and the overcurrent characteristics of RCBOs to BS EN 61009													
Rating (A)		6	10	16	20	25	32	40	50	63	80	100	125	In
Z_s (Ω)		1.92	1.15	0.72	0.57	0.46	0.36	0.29	0.23	0.18	0.14	0.11	0.09	11.5/In

NOTE: The circuit loop impedances given in the table should not be exceeded when the conductors are at their normal operating temperature. If the conductors are at a different temperature when tested, the reading should be adjusted accordingly.

Table 41.4 Maximum earth fault loop impedance (Z_s) for fuses, for 5 s disconnection time with U_0 of 230 V (see Regulation 411.4.8).

(a) General purpose (gG) fuses to BS 88–2.2 and BS 88–6							
Rating (A)	6	10	16	20	25	32	40	50
Z_s (Ω)	13.5	7.42	4.18	2.91	2.30	1.84	1.35	1.04
	Rating (A)	63	80	100	125	160	200
	Z_s (Ω)	0.82	0.57	0.42	0.33	0.25	0.19

(b) Fuses to BS 1361							
Rating (A)	5	15	20	30	45	60	80	100
Z_s (Ω)	16.4	5.00	2.80	1.84	0.96	0.70	0.50	0.36

(c) Fuses to BS 3036						
Rating (A)	5	15	20	30	45	60	100
Z_s (Ω)	17.7	5.35	3.83	2.64	1.59	1.12	0.53

(d) Fuses to BS 1362	
Rating (A)	3	13
Z_s (Ω)	23.2	3.83

NOTE: The circuit loop impedances given in the table should not be exceeded when the conductors are at their normal operating temperature. If the conductors are at a different temperature when tested, the reading should be adjusted accordingly.

4 Appendix 4 – Cable data-resistance, impedance and '$R_1 + R_2$' values

Table 4.1 Resistance of copper cables at 20°C.

Conductor nominal cross-sectional area (mm²)	Maximum resistance of copper conductors at 20°C (Ω/km)
0.5	36
0.75	24.5
1	18.1
1.5	12.1
2.5	7.41
4	4.61
6	3.08
10	1.83
16	1.15
25	0.727
35	0.524
50	0.387
70	0.268
95	0.193
120	0.153
150	0.124
185	0.0991
240	0.0754
300	0.0601
400	0.0470
500	0.0366
630	0.0221
1000	0.0176
2000	0.0090

Notes:
Values are for stranded conductors but solid conductors are nearly identical.
Taken from BS 6360: 1991.

Table 4.2 Values of $R_1 + R_2$ for cables using wire for cpc at 20°C.

Conductor nominal cross-sectional area (mm²)	Maximum resistance of copper conductors at 20°C (Ω/km)
1.5	24.2
2.5	14.82
4	9.22
6	6.16
10	3.66
16	2.30
25	1.454
35	1.048
50	0.774
70	0.536
95	0.386
120	0.306
150	Generally at these sizes a cable cpc is not used. If needed, $R_1 + R_2$ can be worked out from Table 4.1
185	
240	
300	
400	
500	
630	
1000	
2000	

Notes:
Values are for stranded conductors but solid conductors are nearly identical.
Taken from BS 6360: 1991.

Table 4.3 Values of $R_1 + R_2$ for twin and earth cables to BS 6004 at 20°C.

Cable size (mm²)	Size of CPC	$R_1 + R_2$ (Ω/km)	50 m value of $R_1 + R_2$ 50 m run (Ω)
1.5	1	30.2	1.51
2.5	1.5	19.51	0.98
4	1.5	16.71	0.84
6	2.5	10.49	0.52
10	4	6.44	0.32
16	6	4.23	0.211

Table 4.4 Resistance data for armoured cables at 20°C (copper conductors).

Nominal CSA of conductor (mm²)	R_1 resistance (Ω/km)	Armour resistance steel wire			
		2-core (Ω/km)	3-core (Ω/km)	4-core (Ω/km)	5-core (Ω/km)
1.5	12.1	10.2	9.5	8.8	8.2
2.5	7.41	8.8	8.2	7.7	6.8
4	4.61	7.9	7.5	6.8	6.2
6	3.08	7.0	6.7	4.3	3.9
10	1.83	6.0	4.0	3.7	3.4
16	1.15	3.7	3.5	3.1	2.2
25	0.727	3.7	2.5	2.3	1.8
35	0.524	2.6	2.3	2.0	1.6
50	0.387	2.3	2.0	1.8	1.1
70	0.268	2.0	1.8	1.2	0.94
95	0.193	1.4	1.3	1.1	—
120	0.153	1.3	1.2	0.76	—
150	0.124	1.2	0.78	0.68	—
185	0.099	0.82	0.71	0.61	—
240	0.075	0.73	0.63	0.54	—
300	0.060	0.67	0.58	0.49	—
400	0.047	0.59	0.52	0.35	—

Note 1: Data is adapted from BS 5467 for XLPE shaped conductors but other conductor shapes and PVC cables have negligible differences.

Table 4.5 $R_1 + R_2$ data for armoured cables at 20°C, copper conductors.

Nominal CSA of conductor (mm²)	$R_1 + R_2$ values using armour steel wire	
	2- core $R_1 + R_2$ (Ω/km)	4- core $R_1 + R_2$ (Ω/km)
1.5	22.3	20.9
2.5	16.21	15.11
4	12.51	11.41
6	10.08	7.38
10	7.83	5.53
16	4.85	4.25
25	4.43	2.93
35	3.12	2.52

Nominal CSA of conductor (mm²)	$R_1 + R_2$ values using armour steel wire	
	2- core $R_1 + R_2$ (Ω/km)	4- core $R_1 + R_2$ (Ω/km)
50	2.69	2.19
70	2.27	1.47
95	1.59	1.29
120	1.45	0.91
150	1.33	0.80
185	0.92	0.71

Note 1: Data is adapted from BS 5467 for XLPE shaped conductors but other conductor shapes and PVC cables have negligible differences.

Table 4.6 Correction factors for temperature of copper and steel.

Temperature of component	Correction factor for copper	Correction factor for steel
20	1.000	1.000
25	1.020	1.025
30	1.039	1.050
35	1.059	1.075
40	1.079	1.100
45	1.098	1.125
50	1.118	1.150
55	1.138	1.175
60	1.157	1.200
65	1.177	1.225
70	1.197	1.250
75	1.216	1.275
80	1.236	1.300
85	1.256	1.325
90	1.275	1.350
95	1.295	1.375
100	1.314	1.400
105	1.334	1.425

Appendix 5 – Fuse I^2t characteristics

The following fuse I^2t characteristics have been adapted from information formerly produced by GEC switchgear.

I^2t Characteristics

With a prospective current
up to 80kA, 0.15 p.f. at 415 Volts

Type T 2-1250 Amp

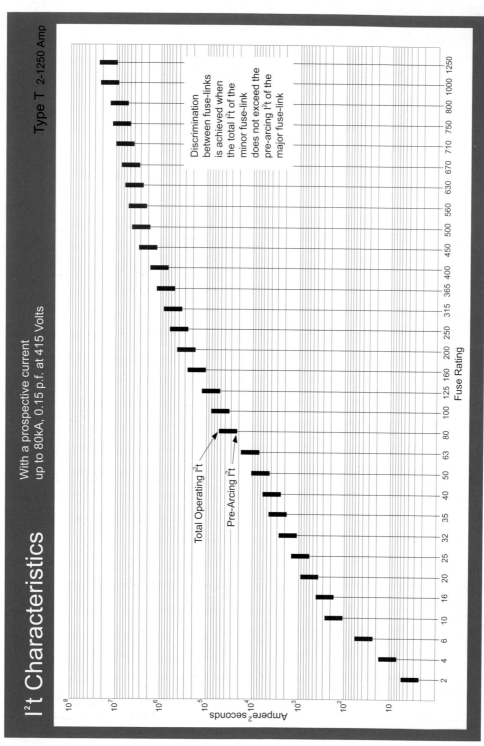

Discrimination
between fuse-links
is achieved when
the total I^2t of the
minor fuse-link
does not exceed the
pre-arcing I^2t of the
major fuse-link

Total Operating I^2t

Pre-Arcing I^2t

Fuse Rating

Ampere² seconds

Index